Cambridge Primary

Hodder Cambridge Primary
Science

Teacher's Pack

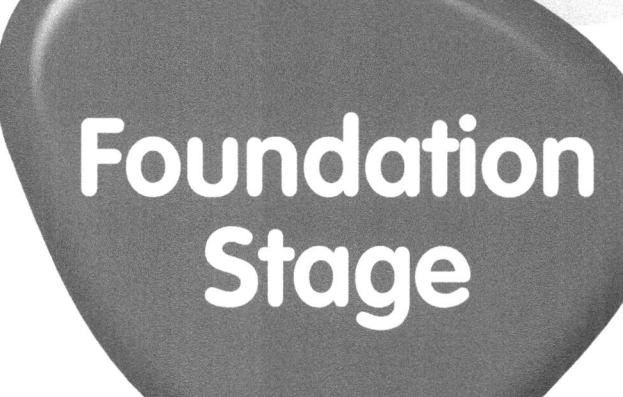
Foundation Stage

Ann and Paul Broadbent

HODDER
EDUCATION
AN HACHETTE UK COMPANY

Acknowledgements

The author and publishers would like to thank Chris Lawson, Science and Early Years Lead, Laurel Avenue Primary School, for her support in planning this material.

The author and publisher would also like to thank Ann and Paul Broadbent for their contribution to pages 6–10 of this Teacher's Pack.

Every effort has been made to trace all copyright holders, but if any have been inadvertently overlooked the Publishers will be pleased to make the necessary arrangements at the first opportunity.

Although every effort has been made to ensure that website addresses are correct at time of going to press, Hodder Education cannot be held responsible for the content of any website mentioned in this book. It is sometimes possible to find a relocated web page by typing in the address of the home page for a website in the URL window of your browser.

Hachette UK's policy is to use papers that are natural, renewable and recyclable products and made from wood grown in sustainable forests. The logging and manufacturing processes are expected to conform to the environmental regulations of the country of origin.

Orders: please contact Bookpoint Ltd, 130 Milton Park, Abingdon, Oxon OX14 4SB. Telephone: (44) 01235 827720. Fax: (44) 01235 400454. Lines are open from 9.00–5.00, Monday to Saturday, with a 24-hour message answering service. You can also order through our website www.hoddereducation.com

© Rosemary Feasey 2018

Published by Hodder Education
An Hachette UK Company
Carmelite House, 50 Victoria Embankment, London EC4Y 0DZ

Impression number   5 4 3 2 1

Year   2020 2019 2018

Cover illustration by Steve Evans

Illustrations by Vian Oelofsen

Typeset in FS Albert 11 pt by Lizette Watkiss

Printed in the United Kingdom

A catalogue record for this title is available from the British Library

978 1 5104 4866 7

Contents

Unit 3 Making sounds

Unit 4 Toys

Unit 5 Food

Unit 6 Dinosaurs

Photocopy masters

Features of each unit

This Teacher's Pack must be used with the *Hodder Cambridge Primary Science Foundation Stage* Activity Books and Story Books.

Icons indicate which components to use in each teaching sequence.

Resources are listed to help teachers prepare and plan for each practical activity.

Background information gives teachers important subject knowledge and pedagogical ideas on the content of each unit.

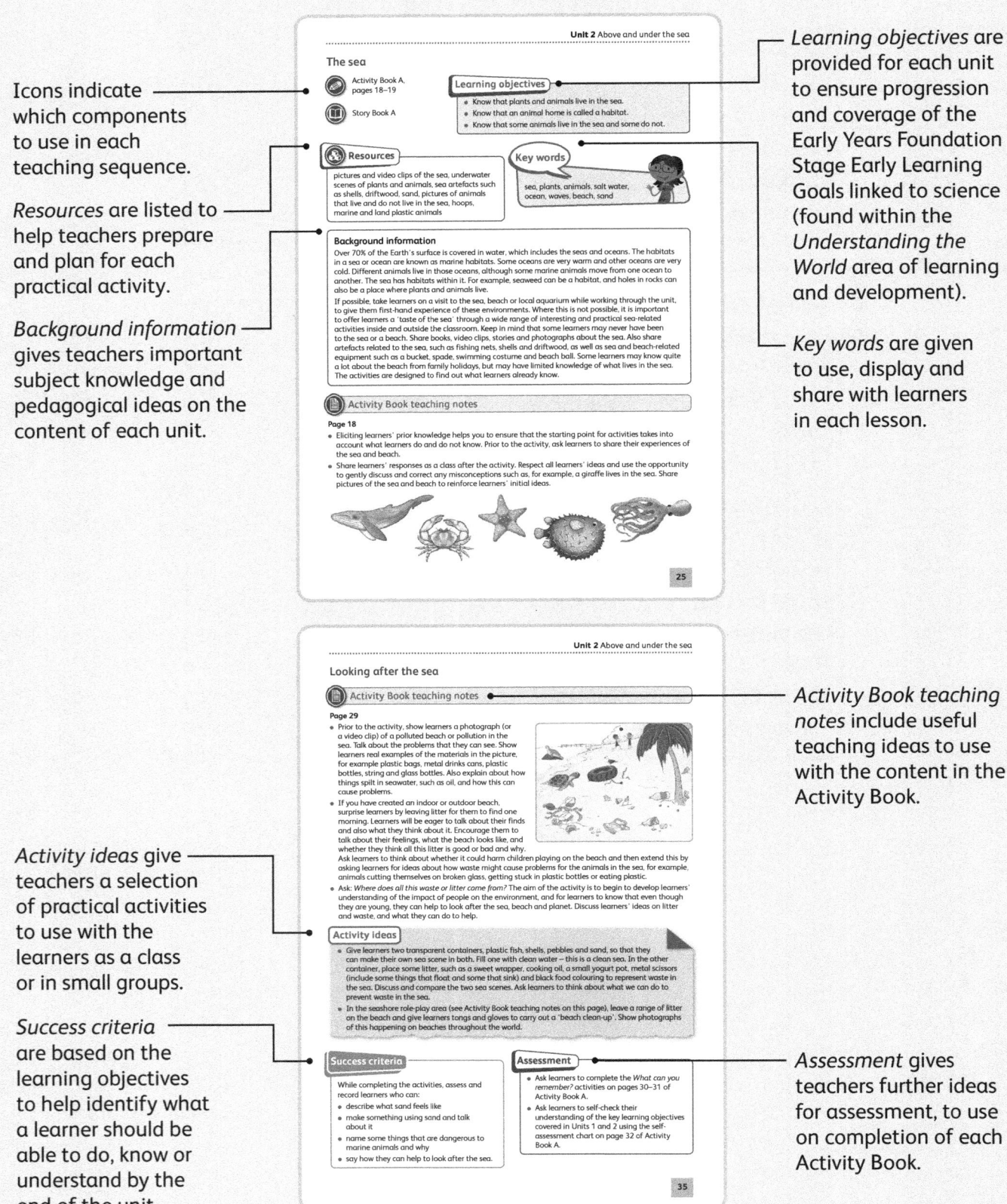

Learning objectives are provided for each unit to ensure progression and coverage of the Early Years Foundation Stage Early Learning Goals linked to science (found within the *Understanding the World* area of learning and development).

Key words are given to use, display and share with learners in each lesson.

Activity Book teaching notes include useful teaching ideas to use with the content in the Activity Book.

Activity ideas give teachers a selection of practical activities to use with the learners as a class or in small groups.

Success criteria are based on the learning objectives to help identify what a learner should be able to do, know or understand by the end of the unit.

Assessment gives teachers further ideas for assessment, to use on completion of each Activity Book.

Introduction

About the series

Hodder Cambridge Primary Science Foundation Stage is written by an experienced author and primary practitioner in Science. The content and progression are based on coverage of the Early Years Foundation Stage Curriculum Framework and provides transition towards Stage 1 of the Cambridge Primary Science Curriculum Framework and *Hodder Cambridge Primary Science Stage 1*.

The Activity Books can be used as stand-alone resources by teachers, parents or carers, covering two units per term. Or the books can be used as part of a complete teaching programme in the classroom for ages four to five, with the content organised into six units over the year (one per half term). Teaching ideas and practical activities are provided in this Teacher's Pack for each of the six units. The units can be completed in any order throughout the school year.

An accompanying Story Book is also included for each Activity Book, containing a science-based story to support the units. The Story Books can be used in the classroom to explore the science themes and can also be shared at home with parents or carers.

Components

Activity Books

There are three Activity Books (A, B and C) – one for each term.

Each Activity Book contains:

- two units, split into popular themes
- coverage of the objectives that underpin the Cambridge Primary Science Stage 1 Curriculum Framework
- full coverage of the Early Years Foundation Stage Early Learning Goals in Science (found within the *Understanding the World* area of learning and development)
- motivating activities and practical tasks, following the mastery approach
- recap activities and a self-assessment chart at the end.

The units in the Activity Books consist of three main types of activity:

Explore: Learners carry out practical activities. Some themes contain pictures, games and diagrams for the teacher and learner to talk about and explore.

Learn: When helpful, a panel with a teaching point that clarifies the concept or skill is included, with a model or image to support understanding.

Apply: Practical and written activities using resources, matching, colouring, drawing and writing, when appropriate, to complete the activities.

Teacher's Pack

The Teacher's Pack contains:

- a complete objectives overview for each unit
- references to the Activity Book pages, the Story Books and the photocopy masters
- a list of objectives at the start of each theme
- teaching notes and practical activity ideas for each theme, to develop understanding, skills and concepts
- background information on each theme, to support the subject knowledge of early years teachers
- key words for new vocabulary to use and display during the activities

- a suggested resource list for each theme
- success criteria at the end of each unit to assist formative and summative assessment
- photocopy masters to support the activities.

 Story Books

There are three Story Books (A, B and C) to support the science concepts in the Activity Books. The Story Books can be read as a class, or in groups, or can be sent home for learners to share with their parents or carers.

Each Story Book contains:

- colourful illustrations and a lively story covering the science themes in the Activity Books
- teacher, parent or carer notes on each page, giving opportunities to talk about the science in the picture or story
- additional activities at the end of the book for the learner to complete.

Structure, scope and sequence

The structure and content of each unit is based on the Early Years Foundation Stage Curriculum Framework and provides transition towards Stage 1 of the Cambridge Primary Science Curriculum Framework.

The stage is divided into six units. Each unit is intended to last approximately half a term. The units can be completed in any order throughout the school year.

Early Years Foundation Stage Curriculum Framework

The Early Years Foundation Stage Curriculum Framework sets out the end-of-year expectations for ages four to five (*Foundation Stage 2/Reception/Kindergarten Upper*), as defined by the following Early Learning Goals in science (found within the *Understanding the World* area of learning development):

> *Understanding the world:* This involves guiding children to make sense of their physical world and their community through opportunities to explore, observe and find out about people, places, technology and the environment.
>
> *People and communities:* Children talk about past and present events in their own lives and in the lives of family members. They know that other children don't always enjoy the same things, and are sensitive to this. They know about similarities and differences between themselves and others, and among families, communities and traditions.
>
> *The world:* Children know about similarities and differences in relation to places, objects, materials and living things. They talk about the features of their own immediate environment and how environments might vary from one another. They make observations of animals and plants and explain why some things occur, and talk about changes.
>
> *Technology:* Children recognise that a range of technology is used in places such as homes and schools. They select and use technology for particular purposes.
>
> This extract is from: *Statutory framework for the Early Years Foundation Stage, Department for Education.*

In science, this involves providing learners with opportunities to:

- explore their environment
- become confident in trying out and testing their own ideas
- develop understanding of materials and their properties
- develop understanding of plants and animals
- become familiar with basic ideas of simple testing
- know how humans, including themselves, can influence their environment.

Hodder Cambridge Primary Science Foundation Stage has full coverage of these Early Learning Goals and careful 'small-step' progression built in to support learners' understanding of skills and concepts.

A mastery approach

Hodder Cambridge Primary Science Foundation Stage uses elements of a mastery approach that are relevant for early years learners:

- Sufficient time is allowed on each theme within a unit for depth of coverage and practice.
- Learners are given opportunities to explore ideas and develop skills in a range of contexts.
- Learners return to ideas in different contexts so that they can develop a deeper and broader understanding.
- Throughout a unit, learners are engaged in practical activities that develop their ability to formulate and carry out simple tests, identify and classify, ask their own questions and use secondary sources to find out information.

Scientific enquiry

Scientific enquiry is embedded into the activities. This helps learners to develop different ideas and ways of working. In each unit, learners experience a range of these types of activities:

> **Observing (over time):** Learners use their senses (where relevant and safe) and observe changes over different periods of time, from immediate (within a few seconds) to an hour, day, week and over a year.
>
> **Pattern seeking:** Learners observe patterns in results, looking for trends and related events.
>
> **Grouping, classifying and identifying:**
> - **Grouping:** Learners sort according to similar observable features or behaviours, such as size, shape and number of legs.
> - **Classifying:** Learners sort according to a scientific grouping, such as type of animal or material. It helps to order a group of things based mainly on similarities rather than differences.
> - **Identifying:** Learners give a scientific name to something, such as *petal*, *seed*, *plastic* and *wood*.
>
> **Comparative testing:** Learners compare one or more objects, but not in a fair test context.
>
> **Researching using secondary sources:** Learners find information using books, video clips, the internet, and leaflets or by asking an expert.

Scientific language

Discussions and talk using accurate scientific language is very important in the classroom for sharing ideas, addressing misconceptions and developing reasoning skills. Key words are given in each unit. Teachers need to plan the introduction of this new vocabulary into the activities and provide opportunities for learners to rehearse and use the words on a regular basis.

Assessment

Formative assessment

On-going, formative assessment is central to teaching and learning in science, particularly in ensuring that learners have achieved more than a superficial understanding, and to avoid moving on too quickly.

Formative assessment should be used to inform the next steps in learning and may influence changes in planning and therefore the next lessons. Formative assessment is a cycle, finding out what learners know, moving learning forward, finding out how that learning has changed (what they know now) and planning next steps. Where you find that learners are still unsure, stop and take time to revisit an idea or skill, change the activity or context and then move onto new learning when learning is secure. Assessment is about you (and the learners) continually reflecting on learning and ensuring that teaching is in line with learning.

Throughout each unit, there are continual opportunities for assessment. As the teacher, you will probe conceptual and procedural understanding through questioning and observation as you model and teach. The way learners respond to the modelling and teaching provides you with valuable information on what to spend a little more time on and what to move through quickly, as well as information on individual needs.

Learning objectives and success criteria

A learning objectives overview is provided at the beginning of each unit in the Teacher's Pack. A list of learning objectives is also given at the start of each theme and suggestions for success criteria are given at the end of each theme. The success criteria are used to help assess the outcome of the learning that has taken place. They are, in effect, what the successful learning will 'look' like once the learning objectives have been met. At the end of each lesson, ask learners to reflect on what they have learned and check their understanding against the success criteria. The self-assessment chart at the end of each Activity Book includes the key objectives and is designed to give learners the opportunity to demonstrate what they know and the concepts they have mastered.

Practical resources

Visual representations, manipulative and concrete resources are hugely important in helping learners to develop a conceptual understanding of what they are learning.

The following general resources are suggested to support the teaching activities and Activity Books for each unit:

- Pictures, photographs and posters
- Artefacts, such as animal teeth
- Video clips or sound recordings
- Small-world toys, such as plastic animals
- Soft toys, for example, animals
- Natural materials
- Junk modelling materials
- Craft materials
- Information books
- Water trays: Indoors and outdoors, and water toys
- Sand trays: Indoors and outdoors, and sand toys
- Measuring equipment for non-standard measurements, for example, plastic cubes and string
- Magnifying glasses
- Plastic transparent pots with lids
- Large plastic hoops.

There are other resources in an early years classroom that can be used to enrich science teaching and learning. The early years environment often includes specific settings for learners to experience. Opportunities can be taken from these settings to develop science concepts and skills. These settings include:

- sand (dry and wet)
- water
- mark-making to formal writing
- reading or books
- small-world toys
- outdoors
- role play
- construction
- music
- 'home' corner
- information technology (IT).

Unit 1 Animals and us

Learning objectives overview

Themes	Learning objectives	Activity Book pages	Preparation for *Cambridge Primary Science Stage 1*
All about me	Name parts of the body. Know the five senses. Describe how the senses are used.	4–5	1Ep1 1Ep4 1Eo4 1Bh1 1Bh2 1Bh4
Similar and different	Know that humans are animals. Know that animals are living things. Talk about similarities and differences between humans and animals.	6–7	1Ep1 1Eo4 1Bh1 1Bh2 1Ep4
What animals need to live	Know that animals are living things. Say what animals need to stay alive. Compare the needs of humans with different animals.	8–9	1Ep1 1Eo2 1Eo4 1Bp2 1Bh1 1Bh3
Animals and their babies	Know that animals have offspring. Match adult animals to offspring. Know that some animals lay eggs.	10–11	1Eo4 1Bh5
Animals moving	Describe how different animals move. Sort animals into groups according to how they move.	12	1Ep1 1Eo2 1Eo4 1Eo6
Animals hiding	Know what *camouflage* means. Know that some animals use camouflage to hide. Name some animals that use camouflage.	13	1Ep1 1Bp3
Where do animals live?	Know that animals live in different places. Know that an animal home is called a habitat.	14–15	1Eo2 1Eo4 1Eo6 1Bp3 1Bh3
A habitat near me	Describe a local habitat. Name some animals that live in a local habitat. Make a model habitat.	16–17	1Eo4 1Eo6 1Bp3 1Bh3

All about me

 Activity Book A, pages 4–5

 Story Book A

Learning objectives

- Name parts of the body.
- Know the five senses.
- Describe how the senses are used.

 **Resources**

dolls, jigsaw puzzles and posters of the human body, model of a human skeleton, musical instruments and objects, lidded pots/boxes, different foods to test senses, different textured materials, fine to coarse sandpaper, plastic bottles, coloured water and cellophane, smell pots

Key words

head, hair, arm, hand, leg, foot, sense, sight, hearing, taste, smell, touch

Background information

This set of activities focuses on body parts. Many learners know basic body parts such as *arm*, *leg* and *head*, but do not know other features such as, for example, *nail*, *wrist*, *knuckle* and *thumb* on the hand. Use a wide range of 'body' vocabulary in these activities. For example, when learners play with dough, ask which part of their hand they use to change the shape of the dough, for example, *thumb* or *palm*. In active games, ask which part of the body learners use to jump, run and throw. This helps to reinforce knowledge of body parts and the corresponding vocabulary.

Introduce the senses by linking body parts to the five senses (sight, taste, hearing, touch and smell). Begin to talk about the idea that like cats, lions, birds, fish, we are animals too, and we have some things in common with other animals, for example, we breathe, move and we have senses.

Activity Book teaching notes

Page 4

- Prior to doing the activity, encourage learners to say and remember parts of the body by playing games, such as 'Simon says', where learners follow instructions, for example: 'Simon says, "Touch your head!"' You could replace the name 'Simon' with a learner's name from your class. This activity will help you to find out which learners are familiar with body part names. Ask learners what the body part does, for example, what the nose does, using the key words *sense* and *smell*.

- Talk about the parts of the body, body part positions and detail, for example: *Our hands have fingers and our fingers have fingernails*. This will ensure that learners are engaged in the process of thinking about what the body looks like.

- Working with, for example, a focus group, draw an outline around one learner's body shape to create a life-size drawing of a body. Ask learners to write labels showing the parts of the body (or provide ready-written labels) and stick these onto the drawing in the correct places. You could use hook and loop tape so that labels can be removed or moved around easily by learners. Display the drawing in the classroom at learner height so that learners can return to the activity and place or replace the labels. Offer additional labels to challenge learners, for example, *ankle*, *wrist* and *neck*.

Page 5

Prior to doing the activity, talk about the senses, and what each sense does. Point to the different parts of the body and the corresponding sense, for example, *ear – hearing*.

Activity ideas

- Use either focus groups or free-flow activities placed around the classroom to provide learners with activities relating to each of the senses. For example:
  - **Hearing:** Create a sound table with different musical instruments and objects that make sounds. Place lidded pots or boxes containing different objects for learners to guess the contents from the sound made when shaken.
  - **Taste:** Provide a range of different foods for learners to try (check for allergies first). Encourage learners to use words such as *sweet, crunchy, chewy* and *sour* as they try the foods.
  - **Touch:** Create a texture wall using different materials fixed to a wall or placed on a table. Include, for example, soft, rough and silky materials. Give learners sheets of sandpaper of different grades, from fine to coarse. Ask learners to describe the sandpaper sheets using the words *rough* and *smooth*.
  - **Sight:** Ask learners to draw their eyes and label the different parts, for example *eyelid, eyebrow* and *eyelashes*. Fill plastic bottles with coloured water, or cover a window with coloured cellophane, for learners to look through. Explore what happens when learners place different coloured pieces of cellophane on top of one another and look through them.
  - **Smell:** Provide 'smell pots' for learners to guess the contents, for example mint, lemon, sugar and coffee. Ask learners to create their own 'smell pot' for friends to guess what is inside.
- Extend the activities by asking learners to think of things that they do which use more than one sense.

Success criteria

While completing the activities, assess and record learners who can:
- name parts of the body
- name the five senses
- point to the parts of the body linked to each sense.

Similar and different

 Activity Book A, pages 6–7

 Story Book A

Learning objectives

- Know that humans are animals.
- Know that animals are living things.
- Talk about similarities and differences between humans and animals.

 **Resources**

pictures of a variety of animals (including humans and birds), magazine pictures of humans and other animals, plastic animals, hoops

Key words

animal, human, same, similar, different, parts of the body, grow, breathe, eat, drink, move, taller, tallest, shorter, shortest

Background information

The concept that humans are also animals is challenging for young learners. The aim of the activities is to support learners in recognising that animals are alive and the similarities and differences between themselves and other animals. This should include comparing body parts and introducing the life processes (evidence that they are alive) in very simple terms:

- To breathe
- To grow
- To move
- To eat and drink.

Other life processes (to reproduce and to excrete) are taught in later years. Use the word *human* when talking about learners, for example: *We are humans, humans are animals.*

 ## Activity Book teaching notes

Page 6

- Prior to the activity, talk about how a human is an animal. Link this to a discussion about how we know humans are alive. Listen to learners' ideas. Introduce the life processes in very simple terms, for example: *We grow, move, eat, drink and breathe. This shows we are alive.* Check that learners understand what each of these life processes mean.

- Talk about how a bird is an animal and a living thing (alive). Discuss that we know this because birds also need to breathe, move, grow, and eat and drink.

- Encourage learners to remember the names of parts of their body and then to think about which body parts the bird has that are the same or similar, for example, legs, head and eyes. Discuss which parts are different, for example, feathers, beak and feet. Show pictures of other birds to highlight that all birds have common features, such as wings and beaks, and that there are similarities and differences between birds and humans.

Page 7

- Build on the language from the previous activity, relating the similarities and differences between the girl and the giraffe. Revise body parts.

- Focus on the height of the giraffe. You could draw a life-size giraffe on the wall or the playground surface, so that learners can stand (or lie) next to it, or work out how many learners it would take to be the same height as the giraffe. Use comparative language, such as *taller, shorter, tallest* and *shortest.*

- Learners could lie on the floor and measure their heights using cut-out paper hands and record the results on a picture of themselves with a sentence: *I am X hands tall.* Measure the giraffe from the above activity using the cut-out hands and compare this measurement to the learners' heights.

Activity ideas

- Give learners pictures of different animals and people to cut out and make a collage. This will help to reinforce the idea that humans are animals too. As learners create their collages, discuss the similarities between themselves and other animals, including body parts and basic needs, for example food, water, and so on.

- Let learners work in pairs. Ask them to show the differences between themselves and a bird through role play, for example, one child is the human while the other is the bird. They could show how they walk and eat, the noises that they make, how they sleep, and so on.

- Ask learners to write (or have scribed) two sentences that show differences or similarities between themselves and a giraffe, for example: *I have a short neck. The giraffe has a long neck.*

- Provide a range of animal pictures or plastic animals for learners to group according to similarities, for example, *has four legs, horns, tails, long necks, beaks, feathers* and *no legs.* Include other ways of sorting, such as *lives in the sea, in trees* or *under rocks,* to find out what learners know about animals.

Success criteria

While completing the activities, assess and record learners who can:
- say why they are an animal
- say how they know an animal is a living thing
- describe how they are similar and different to other animals.

What animals need to live

 Activity Book A, pages 8–9

 Story Book A

Learning objectives

- Know that animals are living things.
- Say what animals need to stay alive.
- Compare the needs of humans with different animals.

Resources

plastic animals, pictures of animals, food, water, pictures of a house, someone sleeping and other items such as a TV, a mobile phone, toys

Key words

animal, living thing, stay alive, air, water, food, shelter

Background information

The life processes (see Teacher's Pack, page 14 for a recap) show that an animal is a living thing. To stay alive, animals (including humans) have these needs:
- Air or oxygen (to breathe)
- Food (to eat)
- Water (to drink)
- Sleep (to rest, some animals do this)
- Shelter (to give protection).

Animals, including humans, need shelter from the environment, (for example, heat, wind and rain), as well as from predators that might eat them. Animals have different ways of making sure that they have the things they need to stay alive. Young learners may say that they need food and water to stay alive, but might not realise that other animals also need food and water.

Air is a difficult concept for young learners as they cannot see it, so keep the explanation simple at this age, for example: *Animals need air to breathe. The air is called oxygen.*

Activity Book teaching notes

Page 8

- Prior to the activity, recap how learners know that they are a living thing, for example, they can move, grow, breathe, eat and drink.
- Discuss with learners what they need to stay alive. Display a variety of pictures of food, a person asleep, a glass of water, houses, a mobile phone, a computer, a television and toys. Ask learners to sort the pictures into 'needed' and 'not needed' to stay alive.

- Talk about the difference between things we like to have and things we actually need to stay alive. Some learners might think that they could not survive without, for example, their toys. Through discussion, develop learners' understanding that they can stay alive without toys, but that they would not survive very long if they did not drink or eat.
- Discuss learners' homes and introduce the word *shelter*. Talk about what it means and what it would it be like to live without any shelter in the cold, rain or extreme heat.
- Introduce the concept of air and how we need it to breathe in order to stay alive.

Page 9

- Ask learners to tell you why the parrot is a living thing, (for example, it moves, eats and drinks, breathes and grows).
- Talk about what the parrot needs to stay alive. Help learners to make the link that just as humans need certain things to stay alive, these needs are similar for other animals, (for example, the parrot needs air, sleep, food, water and shelter).
- Extend the activity to encourage learners to talk about animals that they know and have had experience of. Ask what those animals need to stay alive, for example, what they eat and drink, where they shelter, when they sleep/rest and how they breathe, (for example: *Do they have noses?*).

Activity ideas

- Ask learners to create an 'All about me' book. Include pictures of things humans need to stay alive, such as food, drink and shelter. If possible, encourage learners to bring in from home photographs of their house or of their family eating a meal, and stick these into this book.
- Create a large display about what animals need to stay alive. Stick labels *food, drink, shelter, air* and *sleep* on a wall and add learners' pictures, drawings and photographs around each label. For example, ask learners to sort a selection of foods into 'like' and 'dislike' categories, and create a tally chart of the results. Use the results to create a class pictograph for the food section of the display.
- Talk about other drinks apart from water, for example, milk or juice. Make healthy drinks with learners, for example, flavoured water. Put pieces of apple, pear or orange into water. Talk about which drinks learners prefer and why.
- Invite a visitor into class to show different types of animals, for example, reptiles or insects (ensure that a health and safety risk assessment has been carried out). Learners could ask questions about what the animal needs to stay alive, for example, what it eats and drinks, where it lives, how it breathes and when it sleeps.
- Set up a table-top activity where learners sort pictures or plastic animals into groups according to what they eat, for example, grass, nuts, fruit, meat or meat and plants.
- Learners could do some research to find out about an animal and what it needs to stay alive. They could share their research with the rest of the class or create a picture page showing information.

Success criteria

While completing the activities, assess and record learners who can:
- say why they are a living thing
- say what they need to stay alive
- say what other animals need to stay alive.

Animals and their babies

 Activity Book A,
pages 10–11

 Story Book A

 PCM 1: Animals and their
babies, page 85

Learning objectives

- Know that animals have offspring.
- Match adult animals to offspring.
- Know that some animals lay eggs.

Resources

pictures of animal parents and babies, soft
toys of animals and babies, plastic animals and
babies, eggs, baby photographs of the learners,
video clips of animals hatching from eggs,
ingredients to make animal biscuits, egg-shaped
flap cards, paper plates

Key words

animal, baby, babies, young, eggs,
parent, hatch, similar, different

Background information

This set of activities investigates animals and their offspring. Use learners' own experience of babies
and compare what learners were like as babies by looking at their baby photographs. Talk about
similarities between parents and babies, for example, both having arms, legs, eyes and so on, and also
talk about differences, such as size, eating different things, and not being able to talk or walk. Then
extend this to find out what learners know about other animals and their young, for example, cats and
their kittens, and how they are similar and different. During discussions, encourage learners to use the
correct vocabulary such as *adult, babies, young, similar* and *different,* as well as the correct names for
the animals and their young. In your discussions with learners, include animals that lay eggs, such as
birds, reptiles and amphibians.

Activity Book teaching notes

Page 10

- Prior to starting the activity, share some photographs of learners as babies and discuss the similarities
and differences between them as a baby and now.
- During the activity, focus learners' attention on the similarities between adult and young. For example,
look at features such as a trunk or a beak to help them match up the parent to the baby.
- The frog and tadpole might challenge learners, so discuss what they know about frogs, for example,
where they live (in and near water). Help learners to look for features on the tadpole, such as the tail and
the legs, which might help learners to recognise that tadpoles swim in water.

Page 11

- Most learners will be familiar with eggs, especially at home where they might have seen them used for
cooking. Talk with learners to find out what they know about eggs, show them hen or duck eggs, and
discuss the shape, size and colour.
- Not all learners will know that some animals lay eggs; usually learners know that birds lay eggs, but not
other animals such as insects, reptiles and amphibians. Therefore, prior to completing the activity, show
learners pictures or video clips of a range of animals that lay eggs.

Activity ideas

- Provide opportunities for learners to match the parent with its young, using animal picture cards, soft toys or plastic animals.
- Create a 'Find my parent' hunt around the classroom and outdoors using picture cards of animal parents and babies. Learners, perhaps working in pairs, can hunt for and find the matching animals. You could use *PCM 1: Animals and their babies* (on page 85 of this Teacher's Pack) to create the cards.
- Learners could bake animal biscuits, making a pair, the parent and the baby.
- Give each learner two copies of a paper picture frame. Ask learners to draw and colour a parent animal in one frame and the baby animal in another frame. Place these pictures on display so that all learners can try to match the parent to the baby.
- Create a set of animal cards that learners can sort into those animals that lay eggs and those that do not.
- Give each learner an egg-shaped flap card. Ask them to draw a picture of an adult animal that lays eggs on the front of the card and the baby animal inside. Display the cards on a wall or table top for learners to play 'Guess what is growing inside the egg'.
- Show learners a video clip of an animal hatching from an egg. You could show animals such as crocodiles, birds, snakes and turtles. Stop the video before the baby fully hatches and ask learners if they can work out which baby animal it is.
- Create life cycles for egg-laying animals on paper plates. Show the adult animal, the egg being laid, the egg hatching and the baby animal.
- Extend learning by involving learners in role-playing the adult and the young animal or an animal hatching from an egg.

Success criteria

While completing the activities, assess and record learners who can:
- say that animals have babies
- match an adult animal to its baby
- name some animals that lay eggs.

Animals moving

 Activity Book A, pages 12–13

 Story Book A

 PCM 2: Animals hiding, page 86

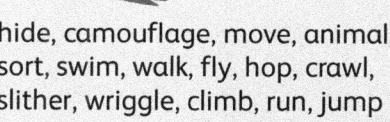
Learning objectives

- Describe how different animals move.
- Sort animals into groups according to how they move.
- Know what *camouflage* means.
- Know that some animals use camouflage to hide.
- Name some animals that use camouflage.

Resources

pictures of animals, pictures of animals moving, pictures of local animals in camouflage, video clips of animals moving and camouflaged, wool, twigs, soil, grass, stones

Key words

hide, camouflage, move, animal, sort, swim, walk, fly, hop, crawl, slither, wriggle, climb, run, jump

Background information

The way an animal moves depends on its features, size, habitat and whether it is a predator or prey. Show learners video clips of different animals moving and ask them to role-play how animals move. This will help them to understand that animals move in different ways and for different reasons. These reasons include: to catch food, to avoid being eaten and to move to shelter.

The animal's features determine how it moves. For example, snakes slither because they do not have legs, the long back legs of frogs help them to jump away from predators and to move for food, and a cheetah has powerful legs to run and catch its prey (food). Some animals stay still or move very slowly and are camouflaged to avoid being seen, so that they are not eaten by a predator or so they can surprise their prey.

Activity Book teaching notes

Page 12

- Prior to the activities, show learners video clips of different animals moving. Discuss how each one moves and develop the associated vocabulary such as *run, hop, slither, jump, swim* and *fly*.
- Develop learners' understanding of how animals move through active sessions where they move like animals, for example, ask learners to move like a penguin, an elephant, a cat, a fish, a snake and a frog. Make sure you choose examples from the different animal groups, such as mammals, reptiles, amphibians, invertebrates (for example, insects), and fish and birds.

Activity ideas

- In active sessions, link animal movement to animals and their habitat, for example, monkeys climb trees and fish swim in water. Ask learners: *How can you show this in your movements?*
- Play learners music related to animals, for example, *The Jungle Book* by Rudyard Kipling, *The Flight of the Bumblebee* by Rimsky-Korsakov and *The Carnival of the Animals* by Camille Saint-Saëns. Ask learners to move like the animals depicted in the music.
- Learners could create a class book called 'Animals and how they move'. Stick pictures, photographs, words and sentences into the book to show what they have learnt about animals and the ways in which they move.

Animals hiding

 Activity Book teaching notes

Page 13

- Prior to completing the activity, show learners pictures of animals from your local environment that use camouflage, such as stick insects, woodlice, moths and snails or larger animals, if appropriate to your country or region. Discuss with learners how they are camouflaged. Ask: *Why are they that colour? Why do they have stripes and patterns?*

- Show learners pictures or video clips of animals that are camouflaged so they cannot be seen by animals that will eat them (predators), or so that the animal that they are trying to find to eat (prey) cannot see them.

- Discuss the shape, colour and features of animals that use camouflage, such as stripes and so on, using pictures. Include a range of different animals such as fish, reptiles and amphibians, as well as mammals and birds.

Activity ideas

- Create a camouflage area in the classroom for plastic or soft toy animals using materials such as twigs, soil, grass and stones.

- Hide toy animals around an outdoor area for learners to find. You could give them a worksheet with a picture of each hidden animal, which they can 'tick' when they have found it. This also helps learners to know when the activity is finished.

- If you have trees or bushes outdoors, challenge learners to draw, paint or make a model animal that is camouflaged to hide in the outside area.

- Learners could draw and cut out the lizard on *PCM 2: Animals hiding* (on page 86 of this Teacher's Pack) and colour it to match one of the backgrounds on a display. Learners could also create their own background against which their lizard can be camouflaged.

Success criteria

While completing the activities, assess and record learners who can:
- talk about how different animals move
- sort animals into groups to show how they move
- say what *camouflage* means
- say why and how some animals use camouflage
- name some animals that use camouflage.

Where do animals live?

 Activity Book A,
pages 14–15

 Story Book A

Learning objectives

- Know that animals live in different places.
- Know that an animal home is called a habitat.

 **Resources**

plastic and soft toy animals, pictures of animals and their habitats, small-world trays, pictures and video clips of different habitats, play dough

Key words

animals, habitat, live, home, sea, desert, jungle, polar areas

Background information

A habitat is a home where animals live, find food, shelter and can raise their young. Learners should explore habitats in the school grounds and locally, and be shown different habitats across the world using pictures and video clips, such as rainforests, deserts and polar regions. Learners should also begin to develop their knowledge of animal characteristics in relation to where they live. For example, camels have large feet to enable them to walk on sand, long eyelashes and nostrils that can open and close to prevent sand getting in them; orangutans have very long arms, so that they can hang onto branches.

 Activity Book teaching notes

Page 14

- Prior to completing the activity, explain to learners that a habitat is where an animal lives. Ask learners: *Where is your habitat?* Discuss where they live, and what their homes provide, for example, shelter, food, water and warmth. Help learners to link the idea that their homes are their habitats, in other words, where they live.

- Provide learners with a range of experiences where they find out about and sort animals that live in different habitats. Discuss with learners what animals look like and their key features, for example, colour and stripes (camouflage), claws (catching prey), wings to fly, thick furry coats to keep them warm, and how these features relate to their habitat.

- Teach learners the names of animals and their habitats. Show them photographs and video clips of different habitats.

Page 15

Remind learners that they are animals, just like birds, fish, snakes and spiders, and that they also live in a habitat. Talk about habitats, what their habitat is and how it is similar or different to other animal habitats. Challenge learners to think about why they do not live in a tree, or in the sea or in the Polar Regions? Posing what might seem to be 'silly questions' can help learners to understand key ideas, for example, you could ask: *Why doesn't a whale live in a tree?*

Activity ideas

- Give learners pictures of two animals such as a polar bear and a penguin to compare. Discuss with learners where these animals live, if they have fur or feathers, their size, what they eat, and so on.
- Extend the activity by giving learners animals from two different habitats, such as the desert and the rainforest, to challenge them to discuss the differences between the habitats and the living things in those areas.
- Provide a range of pictures or plastic animals for learners to sort in different ways, for example, *lives in the sea, in trees* or *under rocks*.
- Set up one or more of the habitats from page 14 in Activity Book A in the role-play area. Place plastic or soft toy animals in these areas – some animals that do and some that do not belong in the habitat. Ask learners to think about this question for each animal: *Do I live here?*
- Ask learners to create habitats outdoors or in the sand, water or other trays for different plastic animals. You could also create 'real' habitats for outdoor animals, for example, mini-ponds for frogs, or logs or stick piles for invertebrates such as woodlice and beetles.
- Take learners into the school grounds so that they can look for animals and their habitats, for example, invertebrates such as snails, spiders and beetles. Discuss why animals live in these habitats. Ask: *Do they live there because it is shady or dry? Is there food for the animals to eat?*
- Learners could make play dough animals to place around the outdoor area or in habitats that they have made.

Success criteria

While completing the activities, assess and record learners who can:

- talk about the places where different animals live
- say what a habitat is.

A habitat near me

 Activity Book A, pages 16–17

 Story Book A

Learning objectives

- Describe a local habitat.
- Name some animals that live in a local habitat.
- Make a model habitat.

Resources

plastic animals, string, feathers, microphone and recording equipment, camera, selection of natural materials such as twigs, grass and leaves

Key words

animals, habitat, live, home

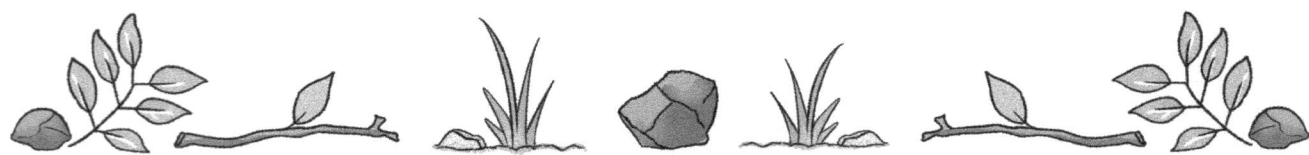

Background information

Where possible, learners should have practical experience of different habitats besides their own school grounds. Take learners out into the local environment to look for animals or invite a visitor to school who can bring in animals for learners to see first-hand. Always make sure that a health and safety risk assessment is carried out first.

When learners are in a different habitat, for example a woodland, talk about the animals and how the habitat, just like learners' own homes, provides shelter, water, food and a place for their young. Help learners to make the connection between habitats and the needs of animals to stay alive.

 ## Activity Book teaching notes

Page 16

Prior to the activity, build on learners' own experiences and knowledge of habitats in their locality and country. Some learners may have visited different habitats (or attractions such as zoos or animal sanctuaries containing habitats) with their parents or carers. Ask learners to talk about these experiences and share photographs. Some families might keep animals, such as goats or hens, which can also provide a focal point for discussion about habitats. Encourage learners to share experiences. Some learners might like to make their own book, poster page or talk into a microphone and record their experiences to replay and listen to.

Page 17

- In this set of activities, the focus is on animals that learners might have come across in the school grounds or the local area, such as birds and invertebrates, such as spiders. This is where being able to draw upon learners' first-hand experiences is important since this set of activities challenges them to apply what they know to make a habitat for an animal. You could take learners on an 'animal safari' around the school grounds to find where animals live, for example, under stones or logs, or on plants. If this is not possible, place pictures of animal habitats and matching animals around the outdoor area.

- Encourage learners to use natural materials from the local environment, such as twigs, to make their model. Provide a range of resources, such as string, feathers and water (to make mud), for learners to use when making the nest.

Activity ideas

- If you have visited a specific habitat with learners, provide opportunities to recreate it either as a role-play area in the classroom or outdoors.
- Ask learners to search for animal habitats in the school grounds or outdoors and take photographs of them to display in the classroom. Ensure that learners understand never to touch or damage a habitat and the reasons why they should not do this.
- Give learners time to look at one another's models and celebrate what they have made and know about habitats.

Success criteria

While completing the activities, assess and record learners who can:
- talk about different habitats and describe a habitat in their country or local area
- say which animals live in a local habitat
- make a model of a habitat and say why they have chosen the materials.

Unit 2 Above and under the sea

Learning objectives overview

Themes	Learning objectives	Activity Book pages	Preparation for *Cambridge Primary Science Stage 1*	
The sea	Know that plants and animals live in the sea. Know that an animal home is called a habitat.	18	1Ep1 1Eo4 1Eo6 1Bp3	
Which animals live in the sea?	Know that some animals live in the sea and some do not.	19	1Ep1 1Eo4 1Bp3	
Sea animals	Name some marine animals. Know some features of marine animals. Know that marine animals live in different places.	20–21	1Eo4 1Ep1 1Bp3 1Bh1	
Comparing sea animals	Name the similarities and differences between marine animals.	22	1Ep1 1Bp3 1Bh1 1Bh4	
Rock pool habitat	Know that a rock pool is a habitat. Know that some plants and animals live in a rock pool and some do not.	23	1Ep1 1Bp1 1Bp3	
Sea turtle	Know that a turtle is an animal that lives in the sea. Name the body parts of the sea turtle. Know the life cycle of a sea turtle.	24–25	1Ep1 1Bp3 1Bh5	
Floating and sinking	Know that some things float and some things sink. Test objects to find out if they float or sink. Make a boat that will float.	26–27	1Ep1 1Ep2 1Ep3 1Ep4 1Eo1 1Eo4 1Eo5	1Cp1 1Cp2 1Cp4
Sand	Describe the properties of sand. Make something using sand.	28	1Ep1 1Ep2 1Ep3 1Ep4 1Eo1 1Eo4 1Eo5	1Cp1 1Cp2 1Cp4
Looking after the sea	Name things that could be dangerous to marine animals. Describe what people can do to look after the sea.	29	1Ep1 1Bp3 1Cp2 1Cp3	

The sea

 Activity Book A,
pages 18–19

 Story Book A

Learning objectives

- Know that plants and animals live in the sea.
- Know that an animal home is called a habitat.
- Know that some animals live in the sea and some do not.

Resources

pictures and video clips of the sea, underwater scenes of plants and animals, sea artefacts such as shells, driftwood, sand, pictures of animals that live and do not live in the sea, hoops, marine and land plastic animals

Key words

sea, plants, animals, salt water, ocean, waves, beach, sand

Background information

Over 70% of the Earth's surface is covered in water, which includes the seas and oceans. The habitats in a sea or ocean are known as marine habitats. Some oceans are very warm and other oceans are very cold. Different animals live in those oceans, although some marine animals move from one ocean to another. The sea has habitats within it. For example, seaweed can be a habitat, and holes in rocks can also be a place where plants and animals live.

If possible, take learners on a visit to the sea, beach or local aquarium while working through the unit, to give them first-hand experience of these environments. Where this is not possible, it is important to offer learners a 'taste of the sea' through a wide range of interesting and practical sea-related activities inside and outside the classroom. Keep in mind that some learners may never have been to the sea or a beach. Share books, video clips, stories and photographs about the sea. Also share artefacts related to the sea, such as fishing nets, shells and driftwood, as well as sea and beach-related equipment such as a bucket, spade, swimming costume and beach ball. Some learners may know quite a lot about the beach from family holidays, but may have limited knowledge of what lives in the sea. The activities are designed to find out what learners already know.

Activity Book teaching notes

Page 18

- Eliciting learners' prior knowledge helps you to ensure that the starting point for activities takes into account what learners do and do not know. Prior to the activity, ask learners to share their experiences of the sea and beach.

- Share learners' responses as a class after the activity. Respect all learners' ideas and use the opportunity to gently discuss and correct any misconceptions such as, for example, a giraffe lives in the sea. Share pictures of the sea and beach to reinforce learners' initial ideas.

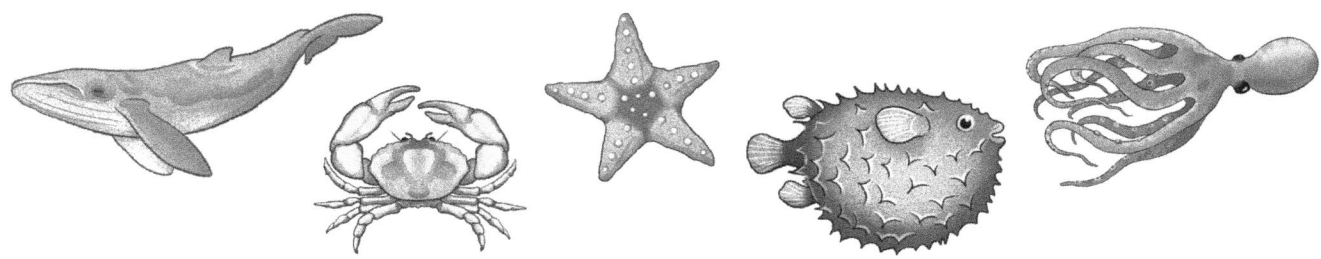

Activity ideas

- Talk about the meaning of a habitat as a home for animals. Discuss with learners why the sea is a habitat. Ask learners about where they live and compare it to the sea habitat where plants and animals live (it provides food, shelter and somewhere for animals to raise their young).
- Start a class Big Book that learners can add to as they find out about the sea.
- Change the water tray into the 'sea' – colour the water blue, add plastic plants and sea animals such as fish, octopuses, whales, and so on.

Which animals live in the sea?

 Activity Book teaching notes

Page 19

- Prior to the activity, talk about animals that live in the sea. Display some pictures of animals for learners to sort into those that do and those that do not live in the sea.
- Discuss the differences, for example their features, between animals that live in the sea and those that do not. Provide some plastic animals, soft toys or puppets (marine and land) to look at and sort during the discussion.

Activity ideas

- Create a role-play area in the classroom or outdoors called 'the seaside'. Include a 'beach' with deck chairs, shells, a beach ball, sand and a rock pool (a water tray) with toy or plastic sea animals. Give learners the opportunity to add artefacts to the area throughout the unit, perhaps using things they have brought from home.
- Place a range of toy animals in the role-play area you created and ask learners to sort them into those that do and do not belong at the seaside.
- Give learners fishing nets to catch animals in a role-play rock pool using the water tray and plastic sea animals.
- Place hoops on the floor and ask learners to sort pictures of a variety of animals into groups, using their own criteria. Then ask them to sort the animals into those that live in the sea and those that do not. Finally, sort the features of animals, for example, *those with a shell* and *those without, those with* and *without fins*, and *those with* and *without claws*.
- If you are able to visit a beach, show learners examples of sea plants such as seaweed. Alternatively, show pictures or video clips of sea plants, so that learners understand that plants as well as animals live in the sea. Maybe try tasting some dried seaweed (found in supermarkets). Check for allergies first.

Success criteria

While completing the activities, assess and record learners who can:
- name some plants and animals that live in the sea
- say what a habitat is
- say why the sea is a habitat.

Sea animals

 Activity Book A, pages 20–21

 Story Book A

Learning objectives

- Name some marine animals.
- Know some features of marine animals.
- Know that marine animals live in different places.

 ## Resources

pictures of the sea and different marine animals, information books about marine animals, toy marine animals

Key words

ocean, angel fish, puffer fish, shark, crab, starfish, octopus, whale, habitat, sea, live, limpet, rocks, shell, fin, fish, scales, skin, gills, tail

Background information

Some animals that live in marine habitats have special ways of breathing (gills), while others can hold their breath under water, coming up for air regularly. Since learners are unable to see what it is like in an ocean first-hand, they will rely on you providing a rich variety of books, video clips, TV programmes and photographs about the sea, as well as sea-related toys and artefacts. The oceans are rich in plants and animals, from very small plankton, which is eaten by different fish, to huge mammals such as whales, which also feed on plankton. Oceans contain different habitats ranging from seaweed where small fish and seahorses live (many seahorses are camouflaged to look like leaves) to holes in rocks where, for example, eels hide, waiting for their latest prey, such as small fish. Some animals, such as giant clams, live on the seabed. Key discussion points are that oceans have different habitats and that these provide sea animals with shelter, food and a place to raise their young. Link this to learners' own homes, where they live (their habitat).

 ## Activity Book teaching notes

Page 20

- Provide learners with a range of pictures or information books about the animals in the activity to help them match a name to each animal.
- After the activity, discuss the features of each animal, for example, the crab has claws and the octopus has eight legs. Ask learners to find similarities and differences between the animals.

Page 21

- Prior to the activity, show learners pictures of different marine animals and discuss their key features and habitats.
- If possible, show a video clip of a limpet and angel fish (or use an information book or a picture). Talk about their body parts and where they live.
- Talk about the differences between the two habitats – the sea and rocks. Ask questions to get learners thinking about why animals choose particular habitats in which to live, for example: *Why doesn't the angel fish live on the rocks?*

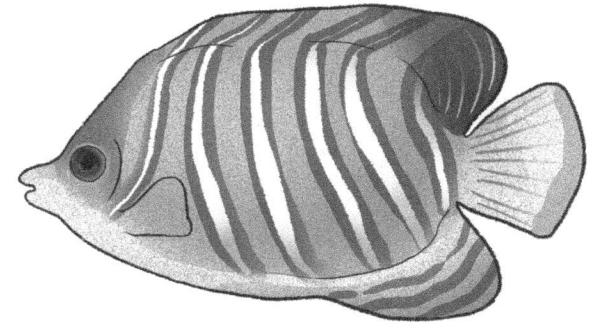

Activity ideas

- Provide picture cards of marine animals and separate name labels. Challenge learners to match the name of the animal to a picture.
- Create 'under the sea' cards, made in the shape of different marine animals, on which learners can write facts (or have them scribed). Display the cards on a wall or table.
- Make and play a game of marine animal 'Snap' with picture cards of animals (include their names). This will help learners to recognise the names and key features of marine animals.
- Use art sessions to make marine animals, for example, an octopus that can hang from the ceiling, or a jellyfish from bubble-wrap. Display the models with information about their key features and habitats.

Success criteria

While completing the activities, assess and record learners who can:

- name some marine animals
- point out some features of marine animals
- talk about the different habitats of marine animals.

Comparing sea animals

 Activity Book A, pages 22–23

 Story Book A

 PCM 3: Sea animals, page 87

Learning objectives

- Name the similarities and differences between marine animals.
- Know that a rock pool is a habitat.
- Know that some plants and animals live in rock pools and some do not.

 Resources

different types of fresh fish, a tray, pictures of marine animals, video clips or pictures of squid and octopuses, photographs, video clips and posters of different habitats in the sea, plastic rock pool animals, hoops

Key words

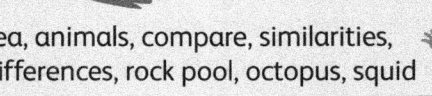 sea, animals, compare, similarities, differences, rock pool, octopus, squid

Background information

Comparing similarities and differences is an important skill. Model the language of comparison for learners, for example, ask: *What is the same or similar? What is different? What are the differences?* When comparing animals, focus on key features and also link to mathematical language such as *smaller* and *bigger*, as well as looking for patterns such as *stripes*, *spots*, and so on.

Rock pools are simply pools of seawater that have collected among the rocks. We can see these pools at low tide, when the sea is out. Plants and animals that live in rock pools must be able to cope with changing conditions. When the tide goes out, the rock pools are left exposed, and as the tide comes in, the pools become part of the sea again. Every day, these plants and animals have to cope with changes in temperature and waves crashing against the rocks, as well as with the salt water being diluted by rainwater. Animals and plants that live in rock pools include sea anemones, crabs, limpets, mussels, shrimps, small fish and starfish.

 Activity Book teaching notes

Page 22

- Show learners pictures of different marine animals and ask them to observe similarities and differences. Support learners in using scientific language by modelling its use. For example:
  - *How is a squid the same as or different to an octopus?*
  - *How is a shark different to a whale?*
  - *What do, for example, angel fish have that starfish do not?*
  You could use *PCM 3: Sea animals* (on page 87 of this Teacher's Pack) for this activity. Learners could pick up any two of the cards and compare similarities and differences between the marine animals.
- Look at a video clip of a squid and octopus swimming in their habitat (or use a picture or a book). Talk about the different features of the animals. Focus on what is the *same* and what is *different*.

Activity ideas

- In small focus groups, show learners some fresh whole fish (different types). Talk about the similarities and differences between the fish. Learners will be fascinated by being able to observe and handle a fresh fish. Ensure learners wash their hands after handling it (and check for allergies before the activity). Place the fish in a tray for easy access. Learners will soon get over being squeamish as their curiosity takes over. Take photographs or video clips of learners handling the fish, so that they can view them later as a stimulus for further discussions.
- Provide shells and different kinds of seaweed for learners to compare, looking at patterns, shape, size, colour, and so on. Remind learners that in the sea, seashells have animals living inside them, and that the seaweed is a plant that lives under the water.

Rock pool habitat

 Activity Book teaching notes

Page 23

- Learners might think that everything is the same under the sea – just water. However, showing learners video clips, posters and photographs of different sea habitats will help them to understand that there are a variety of places where marine animals live, such as on the seabed (lobsters and crabs), among the seaweed (seahorses), in holes in rocks (eels) and in rock pools (limpets and starfish).
- Prior to the activity, talk about the rock pool habitat. Show learners pictures of rock pools and the marine creatures and plants that live in them. Discuss what they look like and what they are called.

- Some learners may have experienced rock pools at the seaside. Ask them to share their experiences with the class. Explain the importance treating marine creatures very gently and returning them to the rock pool after observation.

Activity ideas

- Create a water tray 'rock pool' in the classroom. Include rocks, nets, buckets and plastic animals that would be found in this habitat.
- Place plastic animals that would not be found in the water tray 'rock pool' (see above) and ask learners to sort them into two groups – those that live in a rock pool and those that do not.
- During active games, ask learners to make shapes and movements to represent marine animals and plants, for example, crabs, octopuses, fish, seaweed and sharks. Encourage discussion about why these marine animals and plants move as they do and sequence movements to music to create a simple dance.

Success criteria

While completing the activities, assess and record learners who can:

- say what is similar and different between marine animals
- talk about different habitats in the sea
- name some animals that live in a rock pool.

Sea turtle

 Activity Book A, pages 24–25

 Story Book A

 PCM 4: Sea turtle, page 88

Learning objectives

- Know that a turtle is an animal that lives in the sea.
- Name the body parts of the sea turtle.
- Know the life cycle of a sea turtle.

Resources

pictures or video clips of turtles, toy or plastic sea turtles, photos of the life cycle of turtles and other animals, play dough, clay, junk materials, tessellating fabric or paper shapes

Key words

sea, turtle, flippers, fins, shell, tail, head, eggs, life cycle

Background information

Turtles are reptiles that are found in both fresh and salt water. They have hard shells to protect against predators. They spend most of their lives in water and so have adapted to this habitat – they have flippers and their body is streamlined to move easily through water. Sea turtles rarely leave the sea except to lay eggs in the sand, which they bury to protect from predators. When hatched, the baby turtles (hatchlings) make their way to the sea. When the hatchlings are fully grown, they return as adults to lay their eggs on the beach. It is important to reinforce the idea of a cycle to the learners, for example, young are born, become adults, have young and so the cycle begins again.

Activity Book teaching notes

Page 24

- Before learners complete the activities, they should have opportunities to find out about sea turtles. Use pictures, books and video clips to discuss what they look like and where they live.
- Make comparisons between the features of sea turtles and other sea animals.
- Talk about how sea turtles lay eggs and about their life cycle. Use photographs or books to show learners the different stages in the life cycle. Compare the turtle to other egg-laying animals on the sea and land, such as birds, crocodiles, butterflies and snakes.

Page 25

- Before completing the activity, talk again about the life cycle of the sea turtle (see background information on page 30).
- Demonstrate how to make sea turtles using play dough – ask learners for ideas. Provide a piece of paper with arrows and labels (use the captions on the activity page) for each part of the life cycle models. Photograph each model when completed. Make a display of the learners' life cycle models in a beach scene with sand.

Activity ideas

- Ask learners to make clay turtles and paint them or make model turtles from junk materials.
- Discuss the pattern on turtle shells and create tessellated patterns on 'paper turtle shells'. Give learners pre-cut pieces of paper or fabric that they can stick onto a picture of a turtle. Make sure that they have a photograph to look at so that they know what the pattern looks like on the shell. You could use *PCM 4: Sea turtle* (on page 88 of this Teacher's Pack) for this activity.
- Ask learners to paint a turtle picture and write simple sentences or facts about the turtle. Cut these out and stick them onto the shell of the painted turtle.
- Learners could work with a partner to role-play the life cycle of a turtle, with one learner acting and the other learner narrating.

Success criteria

While completing the activities, assess and record learners who can:
- say where turtles live
- name the parts of the body of a turtle
- describe the life cycle of a turtle.

Floating and sinking

 Activity Book A, pages 26–27

 Story Book A

Learning objectives

- Know that some things float and some things sink.
- Test objects to find out if they float or sink.
- Make a boat that will float.

Resources

water tray, objects that float and sink, variety of junk materials, balloons, pictures of boats

Key words

float, sink, water, under, on top

Background information

The next set of activities link to the sea through exploration of floating and sinking. Learners explore some of the properties of water, and use and apply these experiences to create their own boats. When talking about floating and sinking, focus on asking learners about the material the object is made from, its shape (and also whether they think it has air in it) since these are all factors that contribute to an object floating. The aim is for learners to explore and carry out comparative tests to find out if objects float on water or sink. Give learners the opportunity to choose different objects so that they can predict what will happen and then test their ideas. As learners carry out the test, model the language they should use by talking with them about things that *float* and are *floating* and things that *sink*, are *sinking* and have *sunk*.

Activity Book teaching notes

Page 26

- Prior to the activity, find out what learners know about floating and sinking through a discussion about things that float or sink at the seaside, for example, boats float and stones sink. Introduce the vocabulary *float* and *sink*. Ask learners to share their ideas of where they have seen objects floating and sinking. Some learners might refer to toys in the bath or to when they have been learning to swim, for example, swimming floats. Collect these ideas for a Big Book or a display on floating and sinking.

- As learners explore objects that float and sink, challenge them to use the words *float* and *sink*, and focus on positional language such as *on top* and *under*. Ask learners to think about and share their ideas on why some things float and others sink.

Page 27

- Prior to the activity, discuss the types of material that could be used for a model boat. Listen to learners' ideas about whether the materials will be appropriate, for example, whether they will float or sink. Ensure that learners are allowed to choose materials and build the boat using their own ideas. Some learners will find that their boat sinks straight away. If this happens, discuss the reasons why and encourage learners to try a different idea. For example, ask: *What would happen if you changed the shape of the boat?*

- Give learners a variety of materials to build their boats, including things that float and sink, for example, modelling clay or play dough, paper, tin foil plates, plastic containers, plastic juice cartons, cardboard boxes and wood offcuts.

- Share the results of the boat test with a small group of learners or the rest of the class. Talk about the things that made the boats float and sink. Discuss why some boats were able to carry more plastic cubes than others. Discuss what happened when too many cubes were put on a boat and why this happened.

Activity ideas

- Give learners inflated balloons to put into the water tray. Ask: *Do they float or sink? Why?* Learners will be fascinated by what happens when they try to push the balloon into the water!
- Put a variety of objects (those that will float and those that will sink) into an empty bowl. Ask learners to predict which objects will float and sink. Slowly fill the bowl with water and watch what happens. Discuss the results and whether their predictions were correct.
- Challenge learners to make a floating object sink and a sinking object float. For example, can they do something that would make a small stone float, or if something floats, what could they do or add to make it sink?
- Learners could explore making boats from different materials and different shapes. Provide pictures of different kinds of boats for them to look at, for example yachts, tankers, and so on. Challenge learners to make one of the boats from the pictures.

Success criteria

While completing the activities, assess and record learners who can:

- name some objects that float and sink
- carry out a test and record the results
- use what they know about floating and sinking to make a boat.

Sand

 Activity Book A, pages 28–29

 Story Book A

Learning objectives

- Describe the properties of sand.
- Make something using sand.
- Name things that could be dangerous to marine animals.
- Describe what people can do to look after the sea.

Resources

sand, sand toys, photographs or video clips of a polluted beach or sea, examples of waste such as plastic cartons, metal cans, glass jars, plastic strings, plastic bags, containers, shells, plastic fish, camera, photographs of beach clean-up operations in different countries

Key words

sand, grains, fine, gritty, rough, pours, moulds, shape, wet, dry, animals, litter, dangerous, people, waste

Background information

Experiences with sand at this age promote physical development. For example, large muscle skills develop as learners dig, pour and clean up spills with a brush and dustpan. Eye-hand coordination and small muscle control improve as learners use various sand accessories. Some learners will have visited the beach, while others will have only experienced sand in a more formal setting. So build on and develop learners' understanding by creating an indoor or outdoor beach setting, with access to wet and dry sand and a wide range of accessories for learners to use to explore the sand. This area can be used later to provide a problem-solving setting when 'their beach is polluted by rubbish' that needs to be cleaned and sorted.

Sand is made up of rocks and minerals that have been broken up into very tiny pieces. Exploring the physical properties of sand by playing with it helps learners to discover similarities and differences between wet and dry sand and how sand can be poured.

Pollution of our seas and beaches is a big problem across the world. Learners need to understand the source of the problem (humans throwing away waste and using too many non-recyclable materials) and the possible solutions (using less of materials such as plastic, recycling our waste and not littering beaches). Waste that washes up on beaches consists of many materials including glass, metal and plastic.

We use a lot of plastic in everyday life. It has properties such as being waterproof, hard, flexible and transparent, so it is a very useful material. The problem is disposing of plastic waste – a lot of it finds its way into the sea where it can harm marine animals. Turtles, for example, mistake the plastic for food (for example, jellyfish), and eat it. The plastic then blocks their digestive system and they starve and die. The activities on page 29 of Activity Book A introduce learners to the problem of waste or litter on the beach and in the sea, ready for further study of this in later years.

 Activity Book teaching notes

Page 28

- Give learners the opportunity to explore sand prior to discussion about their own experiences of sand. Ask learners to describe the sand, what they did with it, how it felt and what type of sand it was, for example *wet, dry, fine* or *gritty*. Introduce key language to use when talking about sand.

- In a small group (at the sand tray or outside in the sand pit), explain to learners that they can go and explore the properties of sand – that is, what sand is like using all the senses except taste. For example, look at the sand, watch how it pours, feel the texture of the sand, pour it into different containers and listen to the sound it makes. Provide learners with wet and dry sand for comparison – perhaps in different trays. Encourage learners to feel and describe the sand. Model the language to use when describing the sand, for example: *This is dry sand. It pours easily and feels smooth and soft. This wet sand is gritty, I can mould a shape with it. The sand makes noise like rain when I pour it into the tin.*

- Challenge learners to make things with the sand and, if possible, take photographs of learners' sand creations and stick them onto the activity page.

- After the activity, compare the result of playing with wet and dry sand. Ask: *Which is easier to mould with? Which is easier to pour?*

Activity ideas

- Create a class display about sand by taking photographs of learners' creations in the sand tray. Scribe learners' comments about the properties of sand (and what they can make with it) around the photographs.
- Set challenges for learners in the sand tray, for example, use objects to make tunnels in the sand for vehicles to go through.
- Challenge learners to make marine animals using wet or dry sand.

Looking after the sea

 Activity Book teaching notes

Page 29

- Prior to the activity, show learners a photograph (or a video clip) of a polluted beach or pollution in the sea. Talk about the problems that they can see. Show learners real examples of the materials in the picture, for example plastic bags, metal drinks cans, plastic bottles, string and glass bottles. Also explain about how things spilt in seawater, such as oil, and how this can cause problems.

- If you have created an indoor or outdoor beach, surprise learners by leaving litter for them to find one morning. Learners will be eager to talk about their finds and also what they think about it. Encourage them to talk about their feelings, what the beach looks like, and whether they think all this litter is good or bad and why. Ask learners to think about whether it could harm children playing on the beach and then extend this by asking learners for ideas about how waste might cause problems for the animals in the sea, for example, animals cutting themselves on broken glass, getting stuck in plastic bottles or eating plastic.

- Ask: *Where does all this waste or litter come from?* The aim of the activity is to begin to develop learners' understanding of the impact of people on the environment, and for learners to know that even though they are young, they can help to look after the sea, beach and planet. Discuss learners' ideas on litter and waste, and what they can do to help.

Activity ideas

- Give learners two transparent containers, plastic fish, shells, pebbles and sand, so that they can make their own sea scene in both. Fill one with clean water – this is a clean sea. In the other container, place some litter, such as a sweet wrapper, cooking oil, a small yogurt pot, metal scissors (include some things that float and some that sink) and black food colouring to represent waste in the sea. Discuss and compare the two sea scenes. Ask learners to think about what we can do to prevent waste in the sea.

- In the seashore role-play area (see Activity Book teaching notes on this page), leave a range of litter on the beach and give learners tongs and gloves to carry out a 'beach clean-up'. Show photographs of this happening on beaches throughout the world.

Success criteria

While completing the activities, assess and record learners who can:
- describe what sand feels like
- make something using sand and talk about it
- name some things that are dangerous to marine animals and why
- say how they can help to look after the sea.

Assessment

- Ask learners to complete the *What can you remember?* activities on pages 30–31 of Activity Book A.

- Ask learners to self-check their understanding of the key learning objectives covered in Units 1 and 2 using the self-assessment chart on page 32 of Activity Book A.

Unit 3 Making sounds

Learning objectives overview

Themes	Learning objectives	Activity Book pages	Preparation for *Cambridge Primary Science Stage 1*
What makes a sound?	Know that we use our sense of hearing to listen to sounds. Identify different sounds. Know that the object making a sound is the source of a sound.	4–5	1Ep1 1Eo4 1Bh4 1Ps1 1Ps2
Sorting sounds	Name sounds. Sort sounds using personal preference. Identify the source of a sound.	6–7	1Eo4 1Eo6 1Bh4 1Ps1
Animal sounds	Know that animals make sounds. Know why animals make a sound. Match an animal to a sound.	8–9	1Eo4 1Bh4 1Ps1
Sam's sound hunt	Identify and describe different sounds. Identify objects making a sound. Know that as sound travels from a source, it becomes fainter.	10–11	1Eo4 1Bh4 1Ps1 1Ps3
Musical instruments	Know that musical instruments make different sounds. Name musical instruments. Know how instruments make a sound.	12–13	1Ep1 1Eo4 1Ps1
Loud and soft	Know that sounds can be soft or loud. Know how to make soft and loud sounds.	14–15	1Eo1 1Eo4 1Ps1
Make an instrument	Create a musical instrument. Say how to change the sound on the musical instrument.	16–17	1Ep1 1Ep2 1Eo4 1Ps1

What makes a sound?

 Activity Book B, pages 4–5

 Story Book B

Learning objectives

- Know that we use our sense of hearing to listen to sounds.
- Identify different sounds.
- Know that the object making a sound is the source of a sound.

Resources

objects that make a range of sounds, (for example, musical instruments and toys that make a sound), small containers, small items that make a sound such as paper clips and stones, metal trays, saucepans, beaters, plastic tubing, funnels, water tray

Key words

sound, source of sound, name of sound, senses, ears, hear

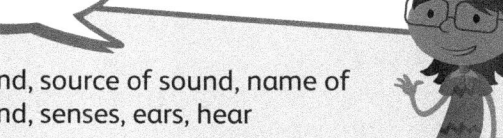

Background information

Sounds are all around us. In this unit, the aim is to raise learners' awareness of their senses and the part of the body that is linked to each sense. While the activities help learners to develop their understanding of individual senses, they also provide learners with opportunities where they use more than one sense at a time, for example, watching and listening to TV, smelling and looking at flowers, or tasting apples and hearing them crunch.

Learners need to know the difference between the source of a sound and the name of a sound. For example, when a bicycle bell makes a ringing sound, the bell is the source of the sound and the ringing is the name given to that sound. Spend some time discussing different examples of this in the classroom, pointing out objects that make a sound (the source of the sound) and naming the sounds they make.

Some learners may need additional support to differentiate between the source of a sound and the sound itself. For example, learners will need an adult to show and talk about the object making the sound, and to differentiate it from the sound itself, for example, by saying: *Make the bell ring. The bell makes a sound, and the sound that the bell makes is a ringing sound.* Ask learners to collect things that make a sound to show you and then to say what sound they can hear.

Activity Book teaching notes

Page 4

- Prior to the activity, talk about the five senses, for example, hearing is one of the senses and we use our ears to hear. Challenge learners to use scientific words when talking about sound and hearing. For example, ask a learner to use a musical instrument to make a sound and ask other learners to point to their ears and say: *I hear things with my ears.*
- Ask learners to point out objects that make a sound in the classroom, school or outside area. Explain that these objects are the source of the sound. Talk about the sounds that these objects make. Ask learners to describe the sounds.

Page 5

- Prior to this activity, offer a range of sound-making activities outdoors that learners can explore and control. For example, use different objects or materials to make *bangs*, *pops* and *drips*.
- Talk about the sounds on the activity page and ask learners for suggestions of the source of the sounds.

Activity ideas

- Make a collection of sound-makers for learners to explore both inside the classroom and outdoors, for example, musical instruments and toys that make sounds.
- Place a range of more unusual objects that can be used to make a sound (such as zips, hook and loop strips, spoons, sandpaper with a scraper, foil and crumpled paper) in an area of the classroom for learners to explore.
- Give learners pairs of containers with objects inside, for example, two pots of stones, feathers, pasta, rice and paper clips (things that make a sound when the containers are shaken). Challenge learners to match containers that make the same sound.
- Place containers made from different materials such as plastic and metal, and plastic piping into the water tray so that learners can explore the sounds that water makes flowing over different materials and into water.
- Provide objects with beaters, for example saucepans and old metal trays, so that learners can explore making different sounds.
- Leave plastic tubing with funnels on each end for learners to talk (or make sounds into) and listen to the sounds their friend makes at the other end of the tube.

Success criteria

While completing the activities, assess and record learners who can:
- say which sense we use to listen to sounds
- name different sounds
- identify different sources of sound.

Sorting sounds

 Activity Book B, pages 6–7

 Story Book B

Learning objectives

- Name sounds.
- Sort sounds using personal preference.
- Identify the source of a sound.

 Resources

range of objects that make sounds, recorded sounds, pictures of sound-makers, fabric sound bag

Key words

sounds, like, dislike, sort, source of sound

Background information

Different objects make different sounds, depending on the materials that they are made from and the size of the vibration making the sound. Develop learners' vocabulary of words that describe sounds, for example, *ring, bang, moo, quack, sizzle, ticking, crying, whistling, gurgle, creak, screech* and *squeak*. Ask learners why they do and do not like different sounds.

Activity Book teaching notes

Page 6

- Talk about sounds learners like and dislike. Identify the source of these sounds, for example: *I don't like the sound of loud banging from this toy drum.* Point out that the *name* of the sound is the *banging* and the *source* is the *drum*.

- Discuss why learners like and dislike certain sounds.

Page 7

- Prior to the activity, explain that some of the pictures show a source of sound and some do not. Focus learners' attention on the objects that make a sound, and that we hear the sound with our ears.

- After the activity, ask learners to name the sounds that the objects make.

- You could play recorded sounds, for example from the internet, and ask learners to listen to the sound, think about the sound, and say what is making the sound, and also what sound is being made.

Activity ideas

- Encourage learners to take ownership of the unit: Ask them to bring in pictures, objects and things that make sounds from home and create a display in the classroom. Involve parents and carers so that they can help extend learners' knowledge by looking for and talking about sources of sound at home.

- Create a class Big Book about sound, where learners can stick pictures and photographs of themselves exploring sound alongside words and their sentences. Use the book to remind learners of key concepts, such as *sources of sounds*, throughout the unit.

- Encourage learners to sort a collection of objects or pictures into sets, 'Sounds I like' and 'Sounds I dislike'. Use a smiley face for 'like' and a picture of someone with their hands over their ears for 'dislike'.

- Let learners work in pairs with a 'fabric sound bag' in which there are sound-makers, for example, paper that makes a crinkly sound, a musical instrument and metal spoons to tap. One learner takes an object out to make a sound, while the other learner faces away so they cannot see the object and tries to work out what it is.

- Create a 'My quiet place' inside and outdoors where learners can sit, in a special seat, to concentrate and listen for sounds, or simply to be still and enjoy being quiet. Give learners time to talk about what they have heard and the sources of the sounds.

Success criteria

While completing the activities, assess and record learners who can:
- identify different sounds
- say why they like or dislike a sound
- name objects that make sounds.

Animal sounds

 Activity Book B, pages 8–9

 Story Book B

Learning objectives

- Know that animals make sounds.
- Know why animals make a sound.
- Match an animal to a sound.

 Resources

pictures of animals, animal sound recordings, animal sound word cards, video clips of different animals, classroom mirror

Key words

animals, sounds, talk

Background information

Animals, including humans, make sounds. They do this to communicate. Humans talk, laugh, cry, shout and think out loud. Other animals communicate to give a warning (stay away this is my territory), to attract a mate, to announce a source of food, to call offspring or to warn of danger.

It is important that learners hear a range of different animal sounds in order to put the activities into context. If possible, play video clips or sound recordings of different animals making sounds. Talk about why animals make sounds.

Activity Book teaching notes

Page 8

- Prior to the activity, ask learners to talk about the sounds humans make, and why they make them. Ask them to think about what animals' sounds they have heard and what they think they mean, for example, the reasons animals make sounds.
- If possible, play learners a sound recording of a bee, an owl, a mouse and a sheep. Make sure learners know that the animals are the sources of the sound and *buzz, hoot, squeak* and *baa* are the names of the sound.

Page 9

- Play learners a sound recording of the sounds made by a duck, lion and snake. Talk about the sound of these animals and the reasons why they might make these sounds.
- Focus on learners remembering the animal names and the sounds that they make, for example, *snake – hiss*.

Activity ideas

- Challenge learners to explore the sounds they can make with their bodies, for example, voice sounds (whistling) and body percussion (clapping, tapping, stamping, and so on). They could watch themselves in the classroom mirror and talk about the parts of the body used and what the sound is like.
- Play learners recordings of different animals. Ask: *What animal is making the sound?* Ask learners to match pictures of animals to the sounds.
- Teach learners words that link to animal sounds using word cards showing a picture of the animal and its sound, for example, *hiss, chirp* and *growl*.
- Play the 'Which animal am I?' game where one learner makes a sound and the other learners have to say which animal it is.

Success criteria

While completing the activities, assess and record learners who:

- name some animal sounds
- say why animals make sounds
- match the animal to the sound it makes.

Sam's sound hunt

 Activity Book B,
pages 10–11

 Story Book B

 PCM 5: Sounds and distance,
page 89

Learning objectives

- Identify and describe different sounds.
- Identify objects making a sound.
- Know that as sound travels from a source it becomes fainter.

 Resources

sound recorder, camera, tablet computer,
blindfold

Key words

sound, fainter, louder, source, near,
far, further, away, distance, hear,
soft, quiet, loud, travels

Background information

Going into the local environment to listen to sounds helps to develop learners' ability to use their five senses, in this case, hearing. When outside, challenge learners to work and think like scientists. They should be quiet and listen, and then use their knowledge and experience to identify sounds. This kind of activity also helps learners apply their sound vocabulary, such as *hear, sound, source, quiet, soft* and *loud*, and to develop new words such as *fainter, travels* and *distance*.

Most sounds that learners hear, such as talking and cars, are made by people, often because they are louder than sounds from the natural world. Develop learners' ability to use their sense of hearing to listen for sounds from the natural world, such as the wind in the trees, birds singing, waves crashing, and so on.

The challenge for learners when experiencing sound getting softer is for them to understand that the sound itself is not getting fainter (the person, for example, banging the drum will be making the same sound at the same 'loudness'), but as the person walks away from the sound, it appears fainter. At this level, it is enough that learners can talk about not hearing the sound as well, or that the sound is softer or fainter.

 Activity Book teaching notes

Page 10

- Ask learners to point out the sources of sound in the picture, which are the: bird, motorbike, bicycle bell, leaves in the palm trees, the water in the puddle, the man behind the stall and the customer talking.
- Talk about the names of the sounds and which sound source they might belong to.

Page 11

- In the outside environment, help learners to focus on using their sense of hearing by asking them to close their eyes and listen. Challenge them to see how many different sounds they can hear in, for example, one minute (which is a long time for young children to stay quiet and still, so you could ask them to lay on the floor with their eyes closed to listen for sounds).

- Talk about whether sounds are near to learners or further away. Learners should begin to use comparative words in science, such as *louder*, *quieter* and *noisier*, at this level. You could use *PCM 5: Sounds and distance* (on page 89 of this Teacher's Pack) to extend the activity.

Activity ideas

- Learners could take photographs of sound sources and record things that make a sound in their local environment, for example, using tablet computers.

- Learners could sing a song or say a nursery rhyme changing how loudly or softly they sing the song, or play clapping games where one learner claps a pattern that includes softer and louder sounds, which the other learner then copies exactly. Both activities require learners to use their sense of hearing and to listen carefully.

- Play the 'Where is the sound?' game. Learners sit in a circle; one learner sits in the middle with a blindfold on or with their hands over their eyes. Another learner is chosen to creep up and gently touch the child on the shoulder. When the seated learner hears the other learner moving towards them, they must point in the direction of the sound. If the learner creeping up gets to the centre without being heard, they take the place in the centre and the game begins again.

- Take learners on a 'natural' sound hunt to listen for wind, leaves rustling and birds singing. Take photographs of what learners can hear for the class 'Sound Big Book' (see page 39 of this Teacher's Pack).

Success criteria

While completing the activities, assess and record learners who can:

- name the sound being made and the object making it
- name sounds from the natural world
- say what happens to a sound as they get further from the source.

Musical instruments

 Activity Book B, pages 12–13

 Story Book B

Learning objectives

- Know that musical instruments make different sounds.
- Name musical instruments.
- Know how instruments make a sound.

 Resources

selection of musical instruments, such as tambourine, guiro, guitar, pan pipes, hoops, castanets, shakers

Key words

sound, musical instrument, pluck, hit, scrape, blow, tap, guitar, drum, pan pipes, tambourine, loud, soft, fast, slow, high, low

Background information

Musical instruments enable learners to try different ways of making sounds. Instruments can be plucked, blown, hit or scraped to make a sound. At this stage, learners do not need to know that a sound is caused by something vibrating, but they do need to understand cause and effect, for example: *If I blow the pan pipes, they make a sound.* Provide learners with plenty of opportunities to play with musical instruments. As learners play with the instruments, encourage discussion about how they make a sound and the different sounds they can make. Make sure learners understand that the instruments are the source of the sound.

Activity Book teaching notes

Page 12

- Prior to the activity, give learners the opportunity to try making sounds with, for example, a tambourine, guiro, guitar and pan pipes.
- Talk about what learners do to make a sound on each instrument. Help learners to link the words *pluck, hit, blow* and *scrape* with how a sound is made.
- Discuss whether there is more than one way to make a sound on each instrument. For example, you can shake and hit a tambourine, and you can hit and scrape a guiro.

Page 13

- Provide a wide selection of instruments for learners to use for this activity. They will need time to experiment with the instruments in order to select four to draw and write about.
- As learners select their instruments, discuss how they make a sound to reinforce the vocabulary *pluck, hit, scrape* and *blow*.

Activity ideas

- Create an orchestra or a musical band practice area where learners can explore and play different musical instruments.
- Encourage role play, for example, learners working together to make a band.
- Provide hoops so that learners can sort the instruments into how the sound is made – plucked, hit, and so on.
- Ask learners to work in pairs. One learner could create a pattern of sound with an instrument and their partner has to remember and repeat this pattern.
- Leave word cards such as *long, happy, sad, fast, slow, high* and *low* and musical instruments on a table. Ask learners to read a word and then make a sound to match it using the instruments.
- Ask learners to create sounds for themes using musical instruments, for example, an 'animal sounds' theme (an elephant trumpeting, a bird singing or a snake hissing) or a 'vehicles' theme (a car, motorbike, aeroplane, tractor or bicycle).

Success criteria

While completing the activities, assess and record learners who can:

- name some musical instruments
- talk about the different sounds made by instruments
- say how an instrument makes a sound.

Loud and soft

 Activity Book B, pages 14–15

 Story Book B

Learning objectives

- Know that sounds can be soft or loud.
- Know how to make soft and loud sounds.

 **Resources**

a range of sound-makers, musical instruments, video clips or sound clips of lions roaring and mice squeaking, pots or boxes with lids, rice, paper clips, marbles, tissue paper, cotton-wool balls, pom-poms, plastic cubes, a balloon

 **Key words**

sound, soft, loud, hard, gentle

Background information

A sound is made when something vibrates. The bigger the vibration, the louder the sound, and the smaller the vibration, the softer (quieter) the sound. If a drum is hit hard, the sound is louder than if the drum is hit gently. If a guitar string is plucked hard, the sound is loud, but if it is plucked gently, the sound is soft, and so on.

For young learners, the key idea is that if they hit, blow or scrape an instrument hard, then the sound will be louder. If they hit, blow or scrape an instrument gently, then the sound will be softer (quieter). It is important that learners are given a wide range of experiences of loud and soft sounds, linked to how they are made, to ensure that they master the idea of how loud and soft sounds are produced.

Ensure that learners understand that they should never play loud sounds close to their ears or they may damage their hearing.

 Activity Book teaching notes

Page 14

- Prior to the activity, give learners a selection of objects for making sounds, such as pots, pans, beaters and musical instruments. Introduce the vocabulary *loud* and *soft* by demonstrating these types of sounds using the objects.
- Ask learners to choose an object and show how to make a loud and a soft sound. Introduce the vocabulary *hard* and *gentle* or *gently*. Model the vocabulary for learners, for example: *If I hit the pan hard, the sound is loud. If I hit the pan gently, the sound is soft.*

Page 15

- If possible, prior to the activities, show learners some video clips or sound recordings of lions roaring and mice squeaking. Compare the sounds using the vocabulary *loud* and *soft*.
- Demonstrate a balloon popping in the classroom. However, be aware that some young learners dislike balloons bursting (it can scare them), so give plenty of warning and allow learners to cover their ears.
- Provide a variety of objects to put into the pots, such as paper clips, plastic cubes and marbles. Include items that will make a soft sound such as tissue paper, cotton-wool balls and small pom-pom balls. Allow learners to test the different items and encourage them to use the vocabulary *loud* and *soft* when discussing the sounds.

Activity ideas

- Use the outdoor area with a focus group and sound-makers, so that learners can explore making very loud sounds and very soft sounds.
- Play the 'Loud and soft' game: Sit in a circle and ask learners to pass an instrument around. Each learner makes either a loud or soft sound and says which sound they are making. Alternatively, show cards with the words *loud* or *soft* and ask learners to make the sound on the card.
- Ask learners to use their voices to make loud and soft sounds. Make patterns of sound that are soft and loud that they copy with their voices.
- Extend learners' understanding of loud and soft sounds by playing different recorded sounds or video clips. Discuss whether the sounds are soft or loud.
- Read stories where learners can make sound effects using different instruments or sound makers, including making loud and soft sounds, for example: *Peace at Last by Jill Murphy (Macmillan Children's Books)*, and *Quiet by Paul Bright (Little Tiger).*

Success criteria

While completing the activities, assess and record learners who can:

● make sounds that are loud and soft

● describe how to make soft and loud sounds.

Make an instrument

 Activity Book B, pages 16–17

 Story Book B

 PCM 6: Sound-maker cards, page 90

Learning objectives

● Create a musical instrument.

● Say how to change the sound on the musical instrument.

 Resources

a range of musical instruments, variety of junk materials for making instruments, such as cans, boxes, containers, plastic bottles, pipe cleaners, elastic bands, beads, stones, rubber bands, small bells, rice, straws and foil plates, camera

Key words

sound-maker, musical instrument, change, sound, drum, rattle, shaker, guitar

Background information

This is a problem-solving activity where learners apply what they know about sound, changing sounds and musical instruments to make their own sound-maker. Problem-solving activities are an opportunity to assess if learners have mastered the key ideas in this unit. By observing and listening to learners, you can assess whether they are able to apply what they have learned. It is also an opportunity for learners to develop their creative abilities in science, for example, in using their imagination in their designs and decoration of the musical instrument.

 Activity Book teaching notes

Page 16

● Prior to the activity, provide a range of musical instruments for learners to look at and try, to give them ideas. Also use *PCM 6: Sound-maker cards* (on page 90 of this Teacher's Pack) for some inspiration.

● As learners draw their designs, talk about how their instrument will work and what will make the sound, and so on.

● Provide plenty of junk materials and allow learners to try various ideas before settling on a final design.

Page 17

● As learners make their instruments, encourage them to think about how the sound will be made and the type of sound. For example, what will they use to fill a shaker or to make the top of a drum?

● When the instruments are finished, have a 'sound celebration' where learners demonstrate their musical instruments and explain how they work. Ask: *How can you use your instrument to make a loud and a soft sound?*

- During this activity, learners apply what they know about sound. This provides an opportunity to assess what they know. Take video clips, photographs or scribe what they say to go in the class Big Book about sound (see page 39) or on a wall display.

Activity ideas

- Encourage learners to apply what they have learned about sound and making different sounds. Challenge them to use their instrument to make as many different sounds as they can, such as quiet and loud sounds, and if they can, high and low sounds, as well as sounds that are quick, sharp and long.
- Some learners will enjoy working in pairs or groups and telling a story for someone else to make the sound effects. Use this activity to encourage learners to explore changing sounds to match a story.
- Encourage learners to play their instruments or tell their story to another class or to parents or carers at the end of the day.

Success criteria

While completing the activities, assess and record learners who can:
- choose materials independently to create a musical instrument
- talk about how to change the sound on their musical instrument.

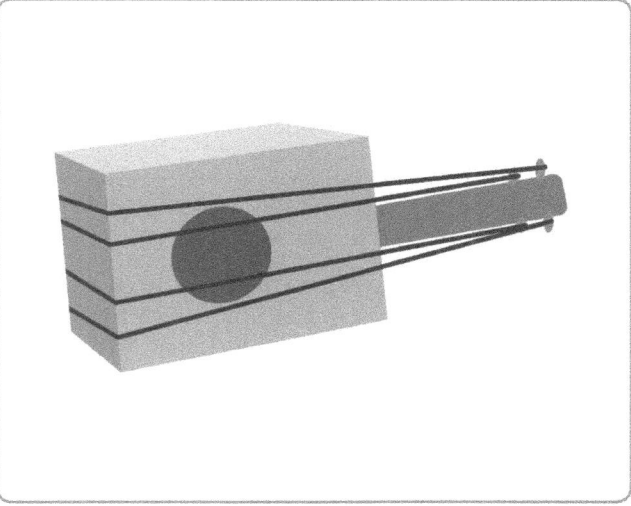

Unit 4 Toys

Learning objectives overview

Themes	Learning objectives	Activity Book pages	Preparation for *Cambridge Primary Science Stage 1*
Sorting toys	Sort toys in different ways. Name some common materials. Sort toys according to the material(s) they are made from.	18–19	1Eo4 1Cp1 1Cp2 1Cp3
Moving toys	Identify toys that can be pushed or pulled. Know that a push and a pull is a force. Carry out a simple comparative test.	20–21	1Ep1 1Eo1 1Eo4 1Pf1 1Pf2 1Pf3
Magnetic toys	Identify magnetic toys. Sort toys according to the material(s) they are made from.	22	1Ep1 1Ep2 1Eo4 1Cp2 1Cp4 1Cp3 1Pf1 1Pf2
Exploring play dough	Talk about the properties of play dough. Identify how pushes and pulls change the shape of play dough.	23	1Ep1 1Ep2 1Eo4 1Cp1 1Cp2 1Pf1
Bubbles	Identify how to make bubbles. Explore how to change the shape of bubbles. Make and test a bubble blower.	24	1Ep1 1Eo1 1Eo4 1Cp1 1Cp2 1Pf1
Shadow puppets	Know how to make a shadow. Make and test a shadow puppet.	25	1Ep1 1Ep2 1Eo1 1Eo6
Build it!	Choose and use different materials for a purpose. Make a construction and test it.	26–27	1Ep4 1Eo6 1Cp2 1Cp3
Make a toy	Apply knowledge of pushes and pulls to design and make a toy. Apply knowledge of materials to design and make a toy.	28–29	1Cp1 1Cp2 1Cp3 1Cp4 1Pf1

Sorting toys

 Activity Book B,
pages 18–19

 Story Book B

Learning objectives

- Sort toys in different ways.
- Name some common materials.
- Sort toys according to the material(s) they are made from.

Resources

hoops, a range of toys of different shapes,
sizes and materials, (for example metal, fabric,
plastic, wood), toys that work in different ways
(for example, battery, wheels)

Key words

toys, sort, set, group, materials,
metal, wood, wooden, plastic,
fabric, battery, wheels, size,
shape, colour, pattern, type,
move, sound, work, texture, hard,
smooth, rough, spiky, bobbly

Background information

Learners are usually very motivated when collections of toys are given to them! As with anything new,
learners will need some time to explore and get to know a selection of different toys – what they do,
how they work, and what they feel and sound like.

Before beginning the activities in this unit, give learners time to familiarise themselves with different
toys. Spend time observing how learners interact with the toys and listen to the language they use. Talk
about toys using vocabulary such as *move, sound, materials, work,* and *battery*. Materials is an area
of science in which learners need constant repetition and reinforcement to master knowledge of the
names of different materials such as wood, metal, plastic and fabric. When sorting the materials, focus
on saying, reading and (for some learners) writing the names of the materials or other sorting criteria.

Activity Book teaching notes

Page 18

- This activity focuses on the range of ways in which learners can sort objects or toys. Begin by giving
  learners the opportunity to sort the toys according to their own criteria. Provide hoops for them to
  sort toys into. Ask them to say how they have sorted the toys, for
  example: *I have put all the toys with wheels together.*

- Some learners will need help with choosing sorting criteria.
  Encourage these learners to sort the toys into materials
  (for example, all the plastic toys together), type (for example,
  all the toy cars together) or how they work (for example,
  all toys that use batteries together).

Page 19

- Prior to the activity, show learners examples of the different
  materials to ensure they understand the sorting criteria. If
  possible, provide some 'real' toys from page 19 for learners
  to look at and discuss.

- Ask learners to name the materials they think each toy is
  made from before completing the activity.

Activity ideas

- Give learners 'pretend' toy boxes labelled with criteria, such as how they work or move, their colour, their size and if they are battery operated. Ask learners to sort selections of toys into the toy boxes.
- Talk with learners about what the toys feel like. Use the word *texture* and introduce the vocabulary to describe properties of materials, for example *soft, hard, smooth, rough, spiky, bobbly*, and so on.
- Ask parents or carers to send in toys from their childhood for learners to compare with their own toys. Talk about similarities and differences in the materials and other features of the toys.
- Give learners hoops or trays labelled with different types of materials, such as *wood, metal, plastic* and *fabric*. Ask learners to sort a selection of toys into the hoops. Ask learners why each toy is made from that material. Challenge them to imagine the toy made from a different material, for example a metal teddy bear or paper bath duck – why wouldn't these materials work?
- Change the role-play area into a toy shop, where learners can pretend to buy, sell, make and repair toys.

Success criteria

While completing the activities, assess and record learners who can:
- sort a set of toys in different ways
- say the material a toy is made from
- sort toys according to the material/s from which they are made.

Moving toys

 Activity Book B, pages 20–21

 Story Book B

Learning objectives

- Identify toys that can be pushed or pulled.
- Know that a push and a pull is a force.
- Carry out a simple comparative test.

 ### Resources

collection of toys that can be pushed or pulled, toy cars, different surfaces (for example grass, sand, soil)

Key words

pull, pulling, pulled, push, pushed, pushing, toys, move, roll, rolling, rolled, test, soil, ground, floor, carpet, force

Background information

Toys provide a very simple introduction to forces for young learners: *Pushes and pulls are forces. Forces can make things move.* The key to the activities is for learners to understand the meaning of *push* and *pull* and that to make something move a *force* is needed. Demonstrate these words using a toy and introduce the word *force*. Talk about other objects that move using a push and a pull, such as a door, a drawer and a pull cord light switch.

Different surfaces affect the way an object moves along it. An object, such as a toy car, moves more quickly on a smooth, flat surface such as a classroom floor than on a bumpy, uneven surface such as grass. The aim at this level is simply to explore the effects of different surfaces on a moving object that is pushed or pulled. Friction and its effect on surfaces are covered in later years.

Activity Book teaching notes

Page 20

- Prior to starting the activity, introduce the words *push, pull* and *force.* Use some toys to demonstrate these words.

- During the activity, focus on the action of making toys move and whether learners used a push or a pull. Help learners to form sentences to communicate what they did, for example: *I pushed the toy car.*

Page 21

- In this activity, learners carry out a simple comparative test where they compare what happens when they push or pull their toys over different surfaces. Prior to starting the activity, ask learners to describe the different surfaces and to predict how they think the toy car will move on these surfaces.

- Support learners in describing their observations, and drawing a conclusion to their test, for example: *It is easier to pull the toy on the classroom floor because it is smooth.*

Activity ideas

- Use moving toys in active sessions. For example, in gymnastics, describe how they move, then recreate those movements (bounce, spin, pop up, roll, and so on).

- Use outdoor toys, such as tricycles and balls, with small groups to try out and talk about how they move and whether a push or a pull is used.

- Prepare a class Big Book, wall or table-top display about toys to which learners can add pictures or photographs of themselves exploring toys and captions.

- Provide learners with different surfaces (including sand and pebbles) and give them access to outdoor surfaces such as soil and tarmac. Encourage them to test a variety of different toys on these surfaces.

Success criteria

While completing the activities, assess and record learners who can:
- sort toys into those that can be pushed or pulled
- give an example of a force
- test the effect of different surfaces on a toy and say the results.

Magnetic toys

 Activity Book B, pages 22–23

 Story Book B

Learning objectives

- Identify magnetic toys.
- Sort toys according to the material(s) from which they are made.
- Talk about the properties of play dough.
- Identify how pushes and pulls change the shape of play dough.

 Resources

range of different types of magnets, plastic bottles with lids, magnetic and non-magnetic objects, paper clips, play dough, 1–6 spinner, camera

Key words

magnetic, non-magnetic, magnets, metal, attract, pick up, move, stretch, squash, squeeze, roll, twist, play dough

Background information

Many toys use magnets, for example, a magnetic fishing game. The key concept in the activities is that magnets attract and repel other magnets, and attract magnetic things made from materials that include iron, nickel and cobalt. At this stage, all learners need to know is that magnets attract some materials and not others, although some learners might be able to link this to metals. Magnets are a non-contact force; they can make something move from a distance without touching the object. This force is called *magnetism*. Pushes and pulls are contact forces, because to make something move, it has to be touched (pushed or pulled).

Play dough offers learners an interesting way to explore the effects of pushes and pulls, and to develop the language to describe the properties of a material (the play dough). Encourage learners to describe how the play dough feels (*smooth*, *squashy* and *flexible*), as well as talk about what they are doing with the play dough, for example: *I am rolling the dough.* Some learners may be able to make the connection between a stretching or twisting action as a 'pull' and a squashing, rolling or squeezing action as a 'push'.

 Activity Book teaching notes

Page 22

- Prior to starting the activity, provide plenty of opportunities for learners to explore magnets informally. Introduce the vocabulary *magnet* and *magnetic*. Explain very simply that if something is magnetic, then the magnet will pick it up.

- Provide magnets, as well as magnetic and non-magnetic objects and toys for learners to try and discuss.

- Provide a selection of toys for the activity – some that are magnetic and some that are not. Encourage learners to describe what they are doing to find out if the toys are magnetic.

Activity ideas

- Set up a treasure hunt in the sand tray containing hidden magnetic and non-magnetic objects. Ask learners to use magnets to find the magnetic objects. Say: *There are X number of magnetic objects in the sand tray. Can you find them all?*
- Ask learners to sort objects into hoops to show which objects are magnetic and which objects are non-magnetic. Some learners may be ready for you to introduce the vocabulary *attract* and *does not attract*.
- Ask learners to do a comparative test to find out how many paper clips different magnets can lift. Provide a variety of magnets and a box of paper clips. Ask learners to record and explain the results in any way they choose.

Exploring play dough

 Activity Book teaching notes

Page 23

- Prior to the activity, ask learners to talk about the properties of the play dough – what it feels like and what they can do with it.
- During the activity, discuss how they use pushes and pulls to stretch, squash, squeeze, roll and twist the play dough, to help them learn the associated language.

Activity ideas

- Learners could make their own play dough and then take and annotate photographs to explain the stages. Talk about the *pushes* and *pulls* used to make the play dough.
- Give learners a card with the words *squash, squeeze, roll, twist* and *stretch*, matching the numbers on a spinner 1–6. Show learners how to make and use a spinner. Then let them spin the spinner and match the number to the action, for example, 1 = stretch, 2 = squash, 3 = squeeze, 4 = roll or 5 = twist, and 6 = they choose what to do to the play dough. If you do not have a spinner, you could use a dice.
- Mathematical language in science is important, so talk about play dough being made *longer, shorter, thinner* and *thicker*.

Success criteria

While completing the activities, assess and record learners who can:

- sort toys into those that are magnetic and those that are non-magnetic
- say the material of the magnetic toys
- describe how play dough looks and feels
- describe how they can use forces (pushes and pulls) to change the shape of play dough.

Bubbles

 Activity Book B, pages 24–25

 Story Book B

 PCM 7: Shadow puppets, page 91

Learning objectives

- Identify how to make bubbles.
- Explore how to change the shape of bubbles.
- Make and test a bubble blower.
- Know how to make a shadow.
- Make and test a shadow puppet.

Resources

bubble mixture, different objects to use to blow bubbles, such as sieves, spoons with holes, tea strainers, pipe cleaners, cardboard tubes and plastic tubes, torches, craft materials, chalk, paint, washing-up liquid, bubble bath, glycerin

Key words

bubbles, blow, air, rainbow, shadow, torch, puppet, opaque, transparent, see-through, bigger, smaller

Background information

Bubbles are made of air trapped inside a hollow 'liquid' ball. Bubbles are transparent and they also reflect colours in the environment. The physics of bubbles is that they always want to be round, so even if they are misshapen to begin, the bubble will always try to become round. Bubbles like to stay wet, so if they touch something that is dry, they will burst. For this reason, you can amaze learners by wetting your hand and then poking your finger or putting your hand into the bubble without it bursting.

Shadows are made when light is blocked by an opaque or translucent object. Learners' bodies are opaque, so they can make shadows using their bodies. Focus discussion on how learners can make shadows using their bodies and how, when they change their shape, their shadow changes and becomes the same shape as the learners.

Activity Book teaching notes

Page 24

- Prior to the activities, blow some bubbles with the learners. Talk about how bubbles are made, how they look and how they behave. Ask learners for ideas to make the bubbles bigger and smaller. Introduce the vocabulary *transparent* or *see-through*, *bigger* and *smaller*.

- Give learners plenty of time to play with bubbles before they design the bubble blower. Some learners may need support, so provide a variety of materials to give them inspiration, for example, using pipe cleaners, cutting holes in paper or plastic plates and cups, making a paper cone and blowing through the thin end or even using their hands.

Activity ideas

- Allow learners to make their own bubble mixtures. Then compare bubbles made by different washing-up liquids or bubble bath, as well as adding glycerine to find out if the bubbles last longer.
- In focus groups, discuss how to make bubbles, change the shape of bubbles, that bubbles are transparent, how the shape of bubbles change when learners blow harder or move the bubble wand, the colours learners can see, (for example, rainbow colours), and so on.
- Ask learners to find the answer to this question: *Are bubbles always round?* Use pipe cleaners to make bubble wands of different shapes, such as circles, squares, triangles, cubes and triangular prisms. Test these shapes. What do learners notice? (All bubbles are round, even when bubbles change shape due to the wind, they all eventually change to be round.)

Shadow puppets

 Activity Book teaching notes

Page 25

- Ask learners to make shadows indoors and outdoors. Focus their attention on what they do to make the shadow. For example, ask: *Where do you point the torch? Where is the shadow? What does it look like (its shape and size)?*
- Introduce the word *opaque* and talk about what it means. Model the language for learners, for example: *My hand is opaque so it makes a shadow when I shine the torch on it.*
- Take learners outdoors and help them to make links with the position of the sun and the position of their shadow.
- You could use *PCM 7: Shadow puppets* (on page 91 of this Teacher's Pack) to give learners ideas for their own puppets.

Activity ideas

- Ask learners to work in pairs and draw around each other's shadow at different times of the day. Ask them to compare similarities between each of the shadows.
- While outdoors, let learners make 'action' shadows, for example, the action of running or kicking a ball. Ask learners to chalk around each other's shadow on the ground. Learners could also paint their 'action' shadows, for example, with rainbow colours or simply by using brushes and water.
- Invite a shadow puppet company to school to perform for the learners. Afterwards, learners could use their own shadow puppets (from page 25 in Activity Book B) to perform a play for the rest of the class.

Success criteria

While completing the activities, assess and record learners who can:
- make bubbles
- say how to change the shape of bubbles
- use what they know to about bubbles to make a blower
- say how to make shadows
- create a shadow puppet.

Build it!

 Activity Book B, pages 26–27

 Story Book B

 PCM 8: Construction challenge, page 92

Learning objectives

- Choose and use different materials for a purpose.
- Make a construction and test it.

 Resources

construction toys and tools, helmets, high-visibility jackets, recyclable materials (such as cardboard boxes, tubes and tins), tape, stones, twigs, planks of wood, boxes, pictures of bridges and towers, materials to build bridges and towers (for example, paper, straw, card, paper, twigs, sticks)

Key words

build, built, construct, construction, size, above, below, next to, tall, taller, tallest, short, shorter, shortest

Background information

Construction is included in this unit not only because it is a popular toy or area of the early years classroom, but also because it helps learners explore and apply ideas about materials, balance, forces and testing. Give learners open-ended construction resources, such as crates, bricks, planks, guttering, cardboard boxes and tubes, so that they can create, problem solve, and change their minds as they build their models. Also offer challenges, for example:

- *How many different materials can you use to build a model?*
- *Can you build the tallest tower / the longest bridge?*
- *Can you build something that moves?*
- *Can you build something to get toy animals or cars across a 'river'?*
- *Can you build a vehicle to carry these plastic bricks?*
- *Can you build a bridge you can walk across?*

To support the construction activities, create indoor and outdoor role-play 'construction sites'. Include toy construction tools, helmets, high-visibility jackets and so on, as adding these also links to developing learners' ideas of careers. Construction links closely to maths (for example, manipulating 3-D shapes, counting, measuring and positional language), so use this opportunity to reinforce relevant vocabulary as learners build their models.

Activity Book teaching notes

Page 26

- Learners love building towers! They could begin using toy bricks and other construction toys – the challenge being to build the tallest and most stable tower. Allow learners to select the construction materials independently. Talk about the pros and cons of each type of construction toy.

- If using large construction toys (such as wooden bricks or crates) remind learners about safety, for example, to check that the areas around them are clear before they start to build.

- Introduce the vocabulary *taller* and *shorter*, modelling examples, for example: *Zara is taller than this tower of blocks.*

Page 27

- Prior to the activity, look at pictures of bridges. Talk about how they are constructed and the basic things needed to build a toy bridge, for example, a supporting structure at each end and a plank.
- Allow learners to select the construction toys independently for their bridge. If their ideas are unsuccessful, support them in choosing a different design or construction material. This is an ideal opportunity to discuss the properties of the materials needed for a bridge. For example, the plank needs to be made from a strong material. In some settings, and depending on the materials used, learners may be able to walk across their bridges.

Activity ideas

- Ask learners to measure their towers by comparing them to their own heights or using non-standard measures such as hand spans.
- Let learners test their towers to find out which one is the strongest and most stable.
- Give learners photographs of different towers from your own country and around the world. Ask them to try and build similar towers using construction toys.
- Challenge learners to build a bridge over some water, for example a tray of water on the floor, or a hole in the soil outside. Make up stories about their bridges, for example, they are over a crocodile infested river or across a steep gorge.
- Challenge learners to build a paper, straw or card bridge that will hold a number of objects, for example, small pebbles.
- Challenge learners to do a simple comparative test of the same bridge or tower design, but using two different materials, to find out which one is stronger and more stable.
- Allow learners to make towers of different materials, such as newspaper, twigs and sticks, spaghetti and marshmallows. Discuss the results.
- Ask an engineer to visit the class to discuss bridge and tower construction, and to show learners photographs of what they have helped to build. They could also bring examples of the construction materials they use and what they wear on a construction site, to show to learners.

Success criteria

While completing the activities, assess and record learners who can:
- choose materials to build towers and bridges independently
- build a tower or bridge successfully
- test their constructions.

Make a toy

 Activity Book B,
pages 28–29

 Story Book B

Learning objectives

- Apply knowledge of pushes and pulls to design and make a toy.
- Apply knowledge of materials to design and make a toy.

Resources

recyclable or junk materials, sticky tape, glue,
string

Key words

materials, recycle, reuse, wood,
plastic, cardboard, paper, fabric,
metal, push, pull

Background information

In this set of activities, learners apply what they have learned about toys to design and make their own toy. Challenge learners to use recyclable materials rather than construction kits. This ensures that they apply mathematical and scientific knowledge about shape and size, as well as the properties of materials, such as strong, flexible and waterproof, to their design.

Activity Book teaching notes

Page 28

Some learners might need to explore the materials prior to putting their design on paper. Give learners time to investigate the materials to find out which is best for their design. Do not worry if their original design does not look like their final model, but encourage them to say why they have made the changes.

Page 29

When learners have completed their toy, take time to support them in reflecting on what they have made and comparing it with their original design. Ask them to show and describe how the construction works, and what it is made from, challenging them to use scientific vocabulary such as *push, pull, squash, material, hard, soft*, and so on.

Activity ideas

- At various stages during the construction process, ask learners why they have chosen certain materials, shapes, and so on.
- Talk with learners about whether their toy has moving parts and how they will make their own toy move, for example, using pushes or pulls.
- Ask learners to make moving toys with, for example, split pins, pop up pages or simple winding mechanisms.
- Celebrate the toys that learners have made by displaying them in a class 'toy shop'.
- Use circle time for learners to show and talk about their toys. Do this over a few days so that learners do not lose concentration when listening to the whole class. Encourage them to comment on and ask questions of the learner who is speaking.

Success criteria

While completing the activities, assess and record learners who can:

- design and make a toy
- describe how their toy moves using, for example, pushes and pulls
- say why they have used certain materials for their toy.

Assessment

- Ask learners to complete the *What can you remember?* activities on pages 30–31 of Activity Book B.
- Ask learners to self-check their understanding of the key learning objectives covered in Units 3 and 4 using the self-assessment chart on page 32 of Activity Book B.

59

Unit 5 Food

Learning objectives overview

Themes	Learning objectives	Activity Book pages	Preparation for *Cambridge Primary Science Stage 1*
Naming and sorting food	Name common foods. Sort food into different groups.	4	1Ep1 1Eo4 1Cp1
Find out about food	Name the five senses. Use the five senses to find out about food. Name common foods.	5	1Eo1 1Eo4 1Bh4 1Cp1
Look inside fruits and vegetables	Use the five senses to find out about food. Describe observations about food. Describe similarities and differences between foods.	6–7	1Ep1 1Eo3 1Eo4 1Eo6 1Bh4
More-healthy and less-healthy foods	Know which foods are more healthy and less healthy. Know that fruits and vegetables are a more-healthy snack. Name common fruits and vegetables.	8–9	1Ep2 1Eo4 1Eo6 1Bh3
Creating with food	Name common fruits and vegetables. Know that fruits and vegetables are a more-healthy snack. Create a fruit dessert.	10–11	1Ep2 1Eo2 1Eo6 1Bh3
Where does food come from?	Know where food comes from. Sort food into different groups.	12–13	1Ep1 1Ep2 1Eo2 1Eo4 1Eo6
Growing food	Know that seeds grow into plants. Know what plants need to grow. Know that some food is grown from seeds. Name the parts of a plant.	14–15	1Ep1 1Ep4 1Eo1 1Eo2 1Eo6 1Bp1 1Bp4 1Bp5
Picnic time!	Apply knowledge to make choices about food. Do a simple comparative test. Draw conclusions using observations.	16–17	1Ep1 1Eo1 1Eo2 1Eo3 1Eo4 1Eo6 1Bh3 1Cp1 1Cp2

Naming and sorting food

 Activity Book C, pages 4–5

 Story Book C

 PCM 9: Fruit and vegetable cards, page 93

Learning objectives

- Name common foods.
- Sort food into different groups.
- Name the five senses.
- Use the five senses to find out about food.

Resources

a range of fruits and vegetables, a range of other foods, chopping board, sharp knife, pictures of people using the different senses, hoops, rubbish bin or plates, pretend food and accessories for the food role-play area, paint, play dough, food and tastes word cards

Key words

food, sort, like, dislike, senses, taste, hearing, sight, touch, smell, fruit, vegetable, juicy, crunchy, hard, soft, sour, moist, sticky, sweet

Background information

Health and safety

During this unit, many of the activities involve learners handling and eating a range of foods. During each activity, ensure that you:

- supervise learners when handling knives to cut food
- follow basic hygiene rules, for example, wash fruits and vegetables, wash learners' hands, and so on
- check for learners who have specific food allergies
- tell parents or carers that learners will taste different foods, and get their consent
- follow your school's health and safety guidelines for working with food in the classroom.

The activities in this unit will help learners to develop good habits towards food, while exploring key science objectives. Give learners plenty of opportunities to look at and explore a variety of foods. This is best done by bringing examples of different types of food into the classroom but, most importantly, please do follow the health and safety advice given above.

For some learners, this may be the first time they hear about the senses. Reinforce the different senses (taste, hearing, sight, touch and smell) by showing pictures of people using their senses. Talk about how we use the senses. Play games that involve the senses and reinforce the vocabulary by asking learners to touch the part of the body when you say a sense, for example, when you say *sight*, learners must touch their eyes.

Activity Book teaching notes

Page 4

- Prior to the activities, have a discussion with learners about their favourite foods.
- Introduce the vocabulary *fruit* and *vegetable*. In a small group, provide a selection of fruits and vegetables, and ask learners to sort these into two hoops labelled 'fruits' and 'vegetables'. Discuss 'trickier' fruits such as the tomato, which some learners may think is a vegetable (it is a fruit).
- If possible, provide examples of the foods in the second activity, for learners to look at, talk about and sort into hoops before they complete the activity.

Find out about food

 Activity Book teaching notes

Page 5

- Prior to the activity, introduce the word *senses* and talk about the five senses. Ask learners to describe how they use their senses. Link the senses to body parts, for example, sight (eyes) and smell (nose). Model examples for learners, for example: *I am using my eyes to see you. I am using my sense of sight.*

- Show learners a selection of fruits and vegetables or other foods, and ask them to describe how they use their senses to find out about the food. Have a sampling session where learners try a selection of new foods.

- During the sampling session, show learners a new food and talk about it before asking them to taste it. In between tasting, learners could drink water to clear their palate. Don't forget to have a bin or plates for learners to discard any food they do not like!

Activity ideas

- Ask learners to sort a selection of food using their own criteria, for example, *like*, *dislike*, *colour* or *shape*. You could use *PCM 9: Fruit and vegetable cards* (on page 93 of this Teacher's Pack) for this activity.

- Create a role-play area in the classroom, focused on food. For example, a restaurant, a café, a kitchen or a picnic area. Provide pretend food and other accessories.

- Make and paint play dough food to add to the role-play area above.

- Have a 'special' snack time where learners can try new foods as well as those they are familiar with.

- Develop the language used to describe food, for example: *juicy, crunchy, hard, soft, sour, peppery, dry, moist, tangy, sticky* and *zingy*. Place the words on cards for learners to match to foods during tasting sessions.

- Play a 'food' memory game. Say: *I went to the shop/market and put in my basket …* Each learner has to say a food, but cannot repeat one that has already been said. This could also be linked to learning the alphabet, for example: *I went to the shop and put in my basket … an **a**pple, a **b**anana and a **c**arrot.*

Success criteria

While completing the activities, assess and record learners who can:

- name some common foods
- sort food into different groups
- name the five senses and associated body parts
- say how they can use their senses to find out about food.

Look inside fruits and vegetables

 Activity Book C, pages 6–7

 Story Book C

Learning objectives

- Use the five senses to find out about food.
- Describe observations about food.
- Describe similarities and differences between foods.

 **Resources**

different fruit and vegetables, chopping board, knives, plates, paint

Key words

similar, different, fruits, vegetables, inside, outside, seeds, peel, flesh, skin, pips, core, stalk, leaves, pith, juice, senses

Background information

This set of activities aims to develop learners' observational skills when looking at food and associated language. Challenge learners to use their senses and focus on similarities and differences between the outside and inside of fruits and vegetables. Model for learners the new vocabulary such as *peel, pith, seeds, juice* and *core*. Provide plenty of examples of fruits and vegetables for learners to compare and discuss, for example, compare the peel on an orange to the skin on a peach.

 Activity Book teaching notes

Page 6

- If possible, provide a selection of the fruits and vegetables on this page. Try and provide two of each type, so that one can be cut open and one remains whole, for comparison. Learners should handle and explore both, and discuss what is the same and what is different.
- Cut open the fruits and vegetables and model the language used to describe them, for example: *Can you see the pips in this apple? What is the orange peel like?*
- Ask learners to use their senses to describe the fruits and vegetables before and after cutting them in half.

Page 7

- Recap the name of the senses and ask learners to describe how they are using their senses when exploring their chosen fruit and vegetable.
- Scribe or provide learners with word cards (for example, *smooth, bumpy, juicy, red, yellow, green, white*), so that they can choose their own vocabulary to complete the activity page.

Activity ideas

- Read a story about fruits and vegetables as a starting point for the activities, such as *Handa's Surprise* by Eileen Browne (Walker Books). Discuss the foods in the book and taste some of of the foods mentioned in the book.
- Make a list of words on the board that describe the similarities and differences between the outside and inside of, for example, an orange: *smooth, bumpy, dry on the outside; juicy, runny, wet and smelly on the inside.*
- Create a set of photographs showing the outside and inside of fruits and vegetables, while learners are engaged in the activities. The learners can then use them as matching activities.
- Use vegetables and fruits in art to print and create regular patterns linked to maths pattern work.
- Use colours from vegetables to paint, for example, beetroot.

Success criteria

While completing the activities, assess and record learners who can:
- name the senses they are using when exploring food
- look carefully at food and describe what they see
- name similarities and differences between foods.

More-healthy and less-healthy foods

 Activity Book C, pages 8–9

 Story Book C

Learning objectives

- Know which foods are more healthy and less healthy.
- Know that fruits and vegetables are a more-healthy snack.
- Name common fruits and vegetables.

Resources

variety of foods, hoops, variety of raw vegetables, chopping board, sharp knife, pictures of more-healthy and less-healthy foods

Key words

snacks, food, fruit, vegetable, choose, balanced, more healthy, less healthy, raw, cooked

Background information

We need food to stay alive. Different foods keep us healthy, prevent us from becoming ill, help us to grow and give us energy. Helping learners to understand that we need a varied and balanced diet is a concept that links all the activities in this unit.

When discussing food, the aim is not to say foods are good or bad, but to help learners develop their understanding that some foods are more or less healthy than others. Introduce the word *balanced*, explaining to learners that this means they should eat lots of different types of foods in moderation, which will include more-healthy and less-healthy foods. Explain that if they eat lots of different kinds of foods, then their body will get what it needs to help them to grow and stay healthy. Reinforce the idea that they should eat plenty of fruits, vegetables, eggs, bread and pasta (and meat or fish if they are not vegetarian), but that they should be aware of (and perhaps limit to 'treat time') consumption of less-healthy foods such as chocolates, biscuits and cakes.

Vegetables are an important part of our diet, as they provide us with vitamins, minerals and fibre. Cooked vegetables are fine, although in the cooking process some nutrients, such as vitamin C, can be lost. Most vegetables can be eaten raw (not all types of beans) and provide a quick and healthy snack for learners.

 Activity Book teaching notes

Page 8

- Prior to starting the activity, provide a range of everyday foods and ask learners to sort them into hoops labelled 'more healthy' and 'less healthy'. Talk about why some foods are more beneficial to the body and why some are not. Point out foods that are best saved for treats and talk about why.
- If possible, show learners examples of foods from the activity page and ask them to talk about why some foods are more healthy and some are less healthy.

Page 9

- Show learners examples of the vegetables from the activity page. Talk about the word *raw*, and its opposite, *cooked*. Ask learners to name the vegetables or match name cards to each vegetable.
- Encourage learners to try a vegetable that they have not tasted raw before. Make it more fun by giving 'well done' stickers when learners try something new.

Activity ideas

- In the role-play area, give learners empty food packets and containers to choose from and use to create pretend 'more-healthy meals'.
- Make sandwiches with learners using locally baked bread. Discuss sandwich fillings and how to make the sandwich balanced, for example: *Can we use a vegetable and cheese?*
- Play the 'swap snack' game: Provide pictures of sweets, cakes or biscuits and ask learners to suggest a 'more-healthy' swap, for example, fruit, vegetables, yoghurt and cheese.
- Learners could make dips to go with their vegetables (for example, a cucumber and yoghurt dip), and take the recipe home to make and share with parents or carers.

Success criteria

While completing the activities, assess and record learners who can:
- say which foods are more healthy and less healthy
- choose a more-healthy snack and say why
- name fruit and vegetables.

Creating with food

 Activity Book C, pages 10–11

 Story Book C

Learning objectives

- Name common fruits and vegetables.
- Know that fruits and vegetables are a more-healthy snack.
- Create a fruit dessert.

 **Resources**

pictures by Italian artist Giuseppe Arcimbaldo, a variety of fruits and vegetables, toothpicks, chopping board, sharp knife, bowls, spoons, yoghurt, cream, ice cream, magazines with pictures of fruit and vegetables, hoops, wooden skewers

Key words

art, picture, head, fruit, vegetables, eyes, nose, ears, mouth, hair, fruit salad, fruit kebab, dessert

Background information

Fruits have many health benefits, for example, citrus fruits and strawberries are rich in vitamin C, which helps the immune system fight illness. Fruits are high in fibre and help the digestive system, and they are also low in calories, which helps to prevent excessive weight gain. Ask learners to think about rainbow colours, such as orange, yellow, red and purple, when choosing fruits and vegetables.

 Activity Book teaching notes

Page 10

- Show learners pictures by Italian artist, Giuseppe Arcimbaldo (1526–1593), who was best known for creating portrait heads made entirely of food, such as fruit and vegetables.

- Give learners the opportunity to explore a variety of fruits and vegetables for their model head. Talk about the shapes needed for each part of the head and encourage learners to choose food independently. If an idea does not work, then support them in trying a different fruit or vegetable.

- This activity works best if learners use a potato or a sweet potato as a head, and foods such as mushrooms, grapes, slices of carrot and pepper and so on, for parts of the face. Explain that the foods can be attached to the potato with toothpicks.

Page 11

- Allow learners to help you prepare a selection of fruits for the dessert, for example, cutting them into bite-size pieces.

- Ask learners to select fruits and begin to make their dessert. They may like to add yoghurt, cream or ice cream to create different layers and patterns. Talk about how the dessert looks, smells and tastes.

Activity ideas

- Learners could cut out pictures of fruit and vegetables from magazines to make their own Arcimbaldo picture.
- Use large hoops into which learners place whole fruit and vegetables to make a face. Photograph the results for a class display.
- Make a fruit salad with learners from the different pieces of fruit already cut up. Ask learners to choose what that they like and encourage them to include new tastes that they have tried.
- Let learners create their own fruit kebabs using wooden skewers. Challenge learners to make a pattern with their fruits, for example, a red-green-red-green fruit or round-square-round piece of fruit.

Success criteria

While completing the activities, assess and record learners who can:

- name common fruits and vegetables
- understand that fruits and vegetables are a more-healthy food
- design and make a fruit dessert.

Where does food come from?

 Activity Book C, pages 12–13

 Story Book C

Learning objectives

- Know where food comes from.
- Sort food into different groups.

 **Resources**

variety of foods, pictures of a variety of foods, hoops, plastic, ceramic or paper plates, play dough, old magazines, vegetable and animal picture cards

Key words

meat, fish, fruit, vegetables, animals, plants, eggs, cheese, bread

Background information

Not all learners know where their food comes from. They may not have thought about this or assume that food comes from the packet. Other learners will be involved in planting food or caring for livestock, so it is important that you discuss this with learners and listen to their ideas and identify any possible misconceptions.

At this age, talk about basic local foods such as fruits, vegetables, meat, fish, eggs, cheese and bread. Avoid foods that have gone through numerous processing stages, which makes it difficult to identify their origins, for example, cakes, biscuits and breakfast cereals. When talking about the source of meat, be sensitive to the fact that some learners may not realise that this comes from animals that were once alive and that some learners might be vegetarian or vegan.

Activity Book teaching notes

Page 12

- Prior to the activity, provide learners with a range of foods (but avoid fresh meat and fish) or pictures of food. Talk about where the food comes from. Learners might say the supermarket or market, so encourage them to think about where the food came from before that. Remember to build on what learners already know. Ask learners to sort pictures of foods (or actual foods) into two hoops labelled 'plant' and 'animal'.

- Ask learners whose family plants food or raises livestock to talk about what they know.

Page 13

Alternatively, you could use paper plates for this activity. Learners could draw, use pictures from magazines, or use play dough, to make their favourite meal and place it on the plate. On the rim of the paper plate, learners could write where their food comes from, for example: *eggs – a hen*; *strawberries – a plant*.

Activity ideas

- Place pictures of food on a table. Ask learners to match a vegetable or animal card to each food.

- Invite someone who bakes bread, or someone who sells vegetables or other foods, to visit the classroom. Learners could ask questions to find out about the visitor's produce and where the food comes from, (for example, vegetables from a farm).

- Create a classroom display showing local fruits and vegetables growing. These could include paintings or printings by learners, for example, a fruit tree. Annotate the display with learners' comments about what they know about local produce and where it comes from.

- Send learners 'shopping' to a role-play supermarket with cards that ask them to buy different things, for example, three foods from plants and two foods from animals.

- Go shopping with a small group of learners to local shops or markets for food that will be used in cold or hot cooking activities with the class. Encourage learners to ask questions about the food, for example: *Where do bananas come from?* Talk to the store or stall holder prior to the visit to help them answer the questions linked to learning.

Success criteria

While completing the activities, assess and record learners who can:

- say that food comes from animals or plants
- sort food into the groups 'from animals' and 'from plants'.

Growing food

 Activity Book C, pages 14–15

 Story Book C

Learning objectives

- Know that seeds grow into plants.
- Know what plants need to grow.
- Know that some food is grown from seeds.
- Name the parts of a plant.

 **Resources**

mustard and cress seeds, plant pots, tissues, compost, pipettes, small watering cans, spoons, pictures of vegetables growing in a garden or window box, pictures of plants showing roots, stem and leaves, digital microscope, plastic cubes, camera

Key words

seeds, plants, grow, grown, planted, stem, leaf, leaves, root, roots, food

Background information

The most practical way for learners to understand that some foods come from plants is for them to plant food that they can eat. The quickest and easiest food to grow is mustard and cress. If the conditions are right, mustard and cress seeds germinate in around 24 hours and are ready to crop within 7 days. Learners can observe and record the growth, and harvest their crop using scissors to snip off the stalks and leaves. Mustard and cress can be eaten by itself, but tastes better in a salad or sandwiches (for example, egg sandwiches).

Salad crops such as radishes and different kinds of lettuce are also quick to grow for young learners who can become disheartened if the plants take too long to grow. Teach learners how to plant and look after seeds. When learners water the seeds, give learners pipettes or spoons to use, so that they do not overwater the plants. Learners can also be asked, for example, to give four teaspoons of water to each plant, thereby encouraging measurement in science.

 Activity Book teaching notes

Page 14

- Prior to the activity, talk about foods that can be grown at home either in a garden or in a window box. Show pictures of food growing, such as lettuce, mustard and cress, chilli peppers, and other salad vegetables. Find out if any learners have tasted these foods.

- Talk about what plants need to grow: warmth, water, something to grow in and on (for example, a pot and compost or soil). Show learners the mustard and cress seeds and discuss how long they think the seeds will take to grow.

- In small groups, plant the seeds, following the process in pictures 1 to 5 at the top of page 14 in Activity Book C:
  1. Put some compost or soil into a pot.
  2. Sprinkle in some seeds.
  3. Use water to wet the compost or soil.
  4. Put the pot in a warm place.
  5. The grown seeds.

  Mustard and cress will grow in various mediums, for example wet tissue or sponge. However, these plants grow best in compost or soil.

- As the mustard and cress seeds grow, discuss learners' observations.
- Develop learners' skills in planting and caring for seeds, and introduce language such as *seed, grow, stem, leaves* and *roots.*

Page 15

- Introduce the basic parts of a plant (stem, roots, leaves) by asking learners to stick labels on a picture or poster of a plant. Point out these parts on the cress and mustard plants and other salad vegetables.
- When the mustard and cress is ready to cut, allow learners to carefully remove a few plants from the pot to examine the roots, stem and leaves. If you have a digital microscope, learners could use this to observe the parts of the plant more closely.
- After cutting the mustard and cress, have a tasting session so learners can try each one. Talk about the parts of the plant they are eating (stem and leaves).

Activity ideas

- Give learners access to a camera to take daily photographs of the plants for a 'Growing diary'. Print out the photographs and stick them in the diary along with captions describing the growth, for example: *Today we could see leaves growing.*
- Have fun growing mustard and cress plants in plastic cups with faces on the front so that the cress looks like hair. Or, glue a photograph of the learner making a face on the cup so that it looks like their hair.
- Invite someone who grows plants to visit the classroom to give learners a 'master-class' in sowing and looking after seeds.
- As the plants grow, use plastic cubes placed beside the plants to show their height at different stages. Each time learners measure the plant, use different colour cubes.
- In a tray, place 'clean' soil or compost along with trowels, plant pots, seed packets and seeds for learners to plant.
- Read *Ten Seeds* by Ruth Brown (Anderson Press Ltd.), a story about growing sunflower seeds. It is well-illustrated and uses scientific terms for parts of the plant as it grows.

Success criteria

While completing the activities, assess and record learners who can:

- name some food grown from seeds
- talk about how their seeds grew into plants
- say what the seeds needed to grow
- name the parts of the plant.

Picnic time!

 Activity Book C,
pages 16–17

 Story Book C

 PCM 10: Sandwich test,
page 94

Learning objectives

- Apply knowledge to make choices about food.
- Do a simple comparative test.
- Draw conclusions using observations.

Resources

bread, plastic boxes, paper bags, metal foil,
butter or margarine, selection of sandwich
fillings

Key words

picnic, food, more healthy,
less healthy, test, wrapper, fresh,
plastic, metal, paper, sandwich,
test, results, material

Background information

There are two elements to these final activities in the unit. The first element is engaging learners in
applying what they know about healthy food to find out if their learning is secure. Successfully applying
what they have learned in a new context (in this case a picnic) provides an assessment of mastery. The
second element is developing learners' ability to plan and carry out a simple comparative test, to find
out what makes the best sandwich wrapper.

Activity Book teaching notes

Page 16

- This activity can be more meaningful to learners if they know that they are planning their lunch for a real
  picnic that might happen later in the week, as perhaps a celebration of the work on this unit.
- Ensure that learners are challenged to use their knowledge about healthy choices to put together a picnic
  for themselves that includes different foods, and rainbow colours when choosing fruits and vegetables.
  Also check that learners are able to say whether their food comes from a plant or animal.

Page 17

- Draw on learners' experiences of wrapping food to eat. Ask them to think about why a sandwich needs
  to be in a wrapper, and list their ideas, for example: *Keep it soft, stop it from being squashed, make sure
  water does not get in*, and so on.
- Give learners the different wrappings (plastic box, paper bag and metal foil) to explore. Ask them how
  they could test each material. For example, they could pour water over each one, poke holes into the
  material, pull the material, and so on.
- Talk through learners' ideas for the test. Some learners will need support. Let these learners plan a test
  together in a small group. Scaffold their ideas with suggestions such as wrapping up a piece of bread in
  each material and then finding out if it is waterproof, resists squashing, and how fresh the bread is after
  being left overnight.
- Provide slices of bread or small bread rolls to test. Learners could make a very simple sandwich using
  butter or spread. However, avoid adding complicated fillings. The aim of the activity is to test the
  freshness of the bread after a period of time in each wrapper. Important: Do not allow learners to eat the
  test sandwiches (for food safety and hygiene reasons).

- At this age, learners may offer a range of suggestions. Do work with their ideas, rather than imposing an adult's perspective, as it is important that learners try out their own ideas, even if they do not work. If the latter happens, it will provide an excellent opportunity to think about why something did not work and what else they could try.

- Work with focus groups to carry out this test. Complete the results of the test the next day, after the sandwich has been left overnight. Learners could use *PCM 10: Sandwich test* (on page 94 of this Teacher's Pack) to record their results.

- Discuss the results with learners and ask them to draw a conclusion using the results, for example: *The plastic box was best for keeping the sandwich fresh because the bread is not dry.*

Activity ideas

- Learners could plan their picnic on one day and then make their picnic to eat on another day. Include making a drink, for example, a fruit squash or fruit-flavoured water.
- Learners could role-play a picnic, using stuffed animals and picnic sets in the outdoors area. You could also place a toy barbecue grill in the outside sand pit or toy kitchen.
- Learners could make their picnic lunch from page 16 in Activity Book C (using the best sandwich wrappers from the test on page 17 in Activity Book C). Take the picnic outdoors or into the school hall, spread picnic blankets out and enjoy the food! Learners could send invitations to a family member to bring their own picnic and join them.

Success criteria

While completing the activities, assess and record learners who can:
- identify and choose more-healthy and less-healthy foods for a picnic
- plan a simple test to compare the effectiveness of sandwich wrappers
- draw a conclusion to say which wrapper is the best for a sandwich.

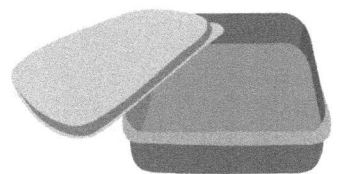
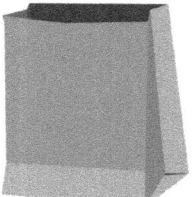

Unit 6 Dinosaurs

Learning objectives overview

Themes	Learning objectives	Activity Book pages	Preparation for *Cambridge Primary Science Stage 1*
Getting to know dinosaurs	Know that dinosaurs were once alive. Know that dinosaurs are extinct. Name parts of the body.	18–19	1Eo4 1Bp2 1Bh2 1Bh1
Comparing dinosaurs	Sort dinosaurs using criteria. Name parts of the body. Identify similarities and differences between dinosaurs.	20–21	1Ep1 1Eo4 1Eo6 1Bh2
Dinosaur habitats	Know that dinosaurs lived in different habitats. Research to find information about a dinosaur habitat. Make a dinosaur habitat.	22–23	1Ep4 1Eo4 1Eo6 1Bp3
What did dinosaurs eat?	Know that dinosaurs ate different things. Identify the features of different dinosaurs.	24–25	1Ep1 1Eo1 1Eo4 1Bp3
How big were the dinosaurs?	Know that dinosaurs were different sizes. Make comparisons between dinosaurs and humans.	26	1Ep2 1Eo4 1Bh1
Fossils	Know what a fossil is. Know what fossils can tell us about dinosaurs.	27	1Ep1 1Eo4 1Bp2
Dinosaur eggs	Know that dinosaurs laid eggs. Talk about the life cycle of a dinosaur.	28	1Eo6 1Bh5 1Bh1
Design a dinosaur!	Apply knowledge of dinosaurs to make a model. Name parts of the body.	29	1Ep4 1Eo2 1Eo4 1Eo6 1Bh2

Getting to know dinosaurs

 Activity Book C,
pages 18–19

 Story Book C

Learning objectives

- Know that dinosaurs were once alive.
- Know that dinosaurs are extinct.
- Name parts of the body.

 ## Resources

pictures of dinosaurs and animals from today, picture of Earth, plastic dinosaurs, toy dinosaurs, pictures of dinosaur fossils and bones, hoops, video clips of dinosaur fossils and bones, tablet computer, model bones, paintbrushes, spoons, trays, art straws, cardboard tubes, papier-mâché

Key words

alive, not alive, extinct, dinosaurs, Earth, died, skeleton, bones, back, tail, arms, feet, teeth, skull, ribs, apatosaurus (ah-pat-uh-sawr-us), pterodactyl (ter-uh-dak-til), triceratops (try-sair-uh-tops), allosaurus (al-uh-sawr-us), tyrannosaurus (tye-ran-uh-sawr-us), palaeontologist (pa-leon-to-logist)

Background information

The word *dinosaur* comes from the Greek language; it means 'terrible lizard'. The dinosaurs lived millions of years ago, but we know about them today because of fossils and bones found in rock. Dinosaurs were vertebrates; they had a skeleton with a backbone. People who study dinosaurs are known as *palaeontologists*.

Although often depicted together in stories, dinosaurs actually lived in three different periods of time. For example, the popular stegosaurus and tyrannosaurus lived at completely different times!

Dinosaurs are split into these periods of time:

- Triassic Period (250–201 million years ago)
- Jurassic Period (201–145 million years ago)
- Cretaceous Period (145–65 million years ago).

Dinosaurs became extinct around 65 million years ago. The exact cause of their extinction is not known, but scientists think that it was due to a large asteroid hitting the Earth or a series of huge volcanic eruptions. Either way, a major event occurred on Earth that wiped out the dinosaurs of the Cretaceous Period.

As a starting point for this unit, you might like to create large dinosaur footprints that lead into and through your classroom, which learners can follow, perhaps to a display about dinosaurs or the book area.

 ## Activity Book teaching notes

Page 18

- Ask learners what they know about dinosaurs and record their ideas.
- Talk about the word *extinct* and ask what learners think it means. Explain to learners that dinosaurs once lived on Earth millions of years ago, (show a picture of Earth), but something happened that killed all the dinosaurs, so now there are none left.
- Talk about animals that are alive today and compare these to the dinosaurs.

Page 19

- Explain that today we know about dinosaurs because people called *palaeontologists* find fossils of their bones.

- Show learners pictures and video clips of dinosaur fossils and bones. Explain to learners that it is rare to find a complete dinosaur skeleton (usually it is only part of a skeleton), so palaeontologists have to put the pieces together like a jigsaw.

- Talk about the parts of the skeleton, such as the skull, teeth, ribs and backbone. Compare dinosaur skeletons with learners' own skeletons. Ask: *Did dinosaurs have ribs and a skull like humans?*

Activity ideas

- Ask learners to sort plastic animals into four hoops labelled 'dinosaurs' and 'not dinosaurs', 'extinct' and 'not extinct'.

- Create a dinosaur 'library' in the book area where learners can sit and look at books, watch clips on a tablet computer, and look at posters about dinosaurs.

- Give learners badges saying, *I am a palaeontologist.* Use the word frequently, for example: *Now, palaeontologists, we are going to find out about dinosaur bones today!*

- Place plaster of Paris or baked dough bones in the sand tray. Give learners paintbrushes, spoons and a tray to find the bones and put them together to see if they can make a dinosaur.

- Place laminated dinosaur bones in a small activity tray for learners to make a dinosaur.

- Make dinosaur skeleton pictures using art straws or pasta tubes. Alternatively, learners could make a floor skeleton using cardboard tubes.

- Let learners make papier-mâché bones to create their own skeletons or part of a huge class dinosaur skeleton.

Success criteria

While completing the activities, assess and record learners who can:

- say how we know dinosaurs were once alive
- say what *extinct* means
- name parts of a dinosaur.

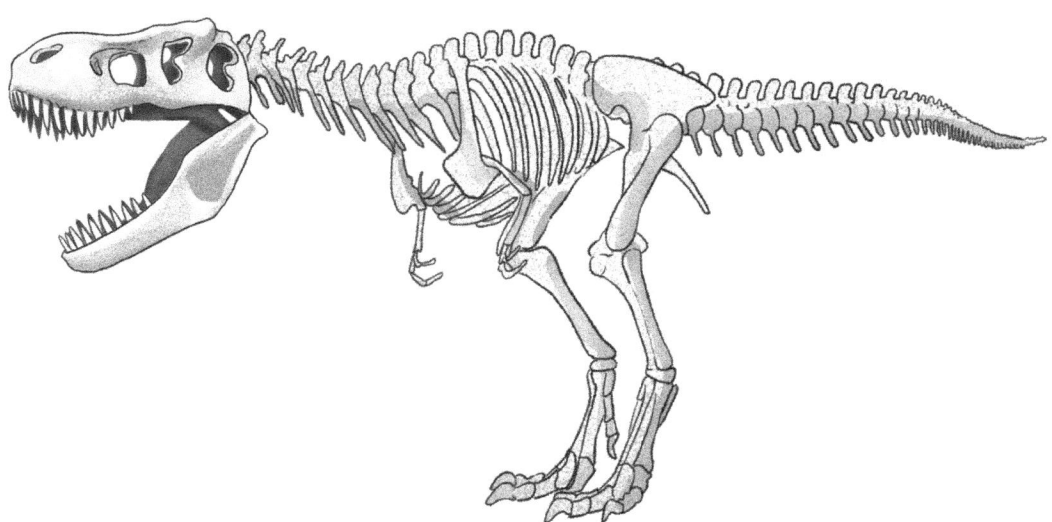

Comparing dinosaurs

 Activity Book C, pages 20–21

 Story Book C

 PCM 11: Dinosaur parts of the body, page 95

Learning objectives

- Sort dinosaurs using criteria.
- Name parts of the body.
- Identify similarities and differences between dinosaurs.

 **Resources**

pictures of dinosaurs and a pterodactyl, dinosaur books, plastic or toy dinosaurs, hoops, paint, dry-wipe pens, chalk

Key words

dinosaur, same, different, similarity, difference, observe, body, parts, eye, spikes, jaw, claws, foot, tail, horn, mouth, teeth, legs, plates, triceratops (try-sair-uh-tops), tyrannosaurus (tye-ran-uh-sawr-us), stegosaurus (steg-uh-sawr-us), apatosaurus (ah-pat-uh-sawr-us), pterodactyl (ter-uh-dak-til), oviraptor (o-vih-rap-tor)

Background information

Making comparisons is important in science; learners need to be able to compare two or more things and talk about their observations in terms of what is similar and what is different. Model the language of comparisons and expect learners to use that language when making comparisons, for example: *Both dinosaurs have a head, but the triceratops's head has horns. Both dinosaurs have a tail, but the apatosaurus's tail is longer.* Learners should be challenged to describe the differences, and not just say they are the same or different.

 Activity Book teaching notes

Page 20

Teach learners the basic parts of the dinosaur's body, such as the foot, tail, horn, mouth, eye and leg. Then move on to different parts, such as spikes, plates, claws and jaw. You could use *PCM 11: Dinosaur parts of the body* (on page 95 of this Teacher's Pack) to make cards, and then ask learners to label pictures of dinosaurs, using these cards as references.

Page 21

- Talk about the fact that the pterodactyl was not a dinosaur. Explain that for some time people thought it was, because it lived around the same time as other dinosaurs.

- Talk about the features of each dinosaur – its body parts, how it moved and what it ate. Explain to learners that some dinosaurs walked on two legs and other dinosaurs walked on four legs. The odd one out is the pterodactyl because it is the only animal that flies. All the other dinosaurs walk or run to move around.

Activity ideas

- Provide plastic dinosaurs and hoops so that learners can sort the dinosaurs into different groups using their own criteria.
- Then ask learners to sort the plastic dinosaurs according to specific criteria, for example, 'has plates', 'has horns' or 'has two legs'.
- Give learners laminated pictures of two similar dinosaurs and ask them to spot the differences. The learners could use a dry wipe pen to mark the body parts that are the same in one colour and those that are different in another colour.
- Ask learners to paint a dinosaur and label the key body parts.
- Provide chalk so that learners can draw dinosaurs in the playground and label key body parts. You could use the cards from *PCM 11: Dinosaur parts of the body* (on page 95 of this Teacher's Pack) for this activity.

Success criteria

While completing the activities, assess and record learners who can:

- compare and sort dinosaurs using criteria
- talk about similarities and differences between dinosaurs
- name parts of a dinosaur's body.

Dinosaur habitats

 Activity Book C, pages 22–23

 Story Book C

Learning objectives

- Know that dinosaurs lived in different habitats.
- Research to find information about a dinosaur habitat.
- Make a dinosaur habitat.

Resources

plastic trays, plastic dinosaurs, mud, grass seed, pebbles, twigs, water, non-fiction books about dinosaurs, posters and video clips showing dinosaurs and dinosaur habitats

Key words

dinosaur, habitat, live, swamp, rainforest, grass, trees, swampy, wet, muddy, water, plants

Background information

Learners might think that all dinosaurs lived at the same time and in the same place. In fact, dinosaurs lived at different times (see background information on page 74) and in different places or parts of the world. Also, within the same time period (for example, the Cretaceous Period), the dinosaurs lived in different habitats, depending on their diet, features and preferences. For example, the stegosaurus ate plants and lived in tropical forests, while rivers and swamps were ideal for the baryonyx, which ate fish.

Dinosaur habitats included the banks of rivers and near the sea where there was vegetation and moisture. They also lived in forests, swamps and grasslands.

Activity Book teaching notes

Page 22

● Introduce the word *habitat*. Ask learners to say what they think it means. Talk about local habitats and compare these to the habitats where dinosaurs lived.

● Show learners pictures and video clips of different dinosaur habitats. Talk about why dinosaurs chose to live in these habitats (for example, for food, to raise young and for shelter).

● 'Dinosaurs' is an excellent unit to develop learners' ability to research information. Provide plenty of resources, such as books, websites, posters and video clips about dinosaurs, for learners to use during the activities.

Page 23

● Prior to the activity, show learners pictures and video clips of dinosaurs in swamp areas. Ask learners to observe what the habitat is like, focusing on words such as *wet, swampy, muddy, sticky, steamy, big ferns* and *plants*.

● Give learners a choice of materials to make their own tray swamp using mud or mixtures such as corn flour and food colouring or bath foam. They could use stones, build mountains from modelling materials, include plants from collage materials, twigs, and so on. Challenge learners to keep referring to the reference material to make the swamps as real as possible for their dinosaurs.

Activity ideas

● Ask learners to create a fact card about dinosaurs, which should include a picture and key information such as height, length, what it ate and its habitat.

● Develop a class dinosaur display and ask learners to add key facts. You could incorporate a photograph of each learner into the display. Learners could then place a caption with a key fact next to their photograph.

● Play the 'Which dinosaur am I?' game. Display a poster or pictures of several dinosaurs. Describe a dinosaur, for example: *I have three horns; I have plates on my back; I am a _____.* Then ask learners to say which dinosaur it is. You could also ask a learner to describe a dinosaur while other learners guess its name.

● Learners could create their own dinosaur museum, using bones and dinosaurs that they have made or using toy ones. They could either create a voice commentary or write labels and captions for their museum.

● Learners could sow grass seed in their swamp soil or mud and wait for it to grow so that they create their own vegetation (do this at the beginning of the unit).

● Provide more than one tray per pair of learners so that they can create different levels for their dinosaurs to move between swamps or from one kind of habitat to another.

Success criteria

While completing the activities, assess and record learners who can:

● talk about the different habitats where dinosaurs lived

● find information about a dinosaur habitat

● apply what they know about dinosaurs to make a model habitat.

What did dinosaurs eat?

 Activity Book C, pages 24–25

 Story Book C

 PCM 12: Dinosaur teeth cards, page 96

Learning objectives

- Know that dinosaurs ate different things.
- Identify the features of different dinosaurs.

Resources

video clips and posters, pictures of different dinosaurs (carnivores, omnivores, herbivores), pictures of dinosaur skulls and teeth, play dough, clay, camera, chalk, sand tray

Key words

dinosaurs, teeth, sharp, flat, meat, plants, ate, food, carnivore, herbivore, omnivore, brachiosaurus (brack-ee-uh-sawr-us), allosaurus (al-uh-sawr-us)

Background information

Most dinosaurs ate plants and the largest dinosaurs, such as the brachiosaurus and apatosaurus, were plant eaters (herbivores) and had flat teeth. They were too big to be able to run and chase food.

Meat eaters such as the tyrannosaurus were carnivores and had sharp teeth and claws. They were smaller than some herbivores and built to run so that they could catch their food. Other carnivores included the velociraptor, which would kill and eat small animals, but also hunted in packs (like wolves today) to enable them to kill larger animals.

Dinosaurs such as the ornithomimus (or-nith-uh-my-mus) and oviraptor (o-vih-rap-tor) were omnivores, as they ate both plants and animals. The skulls of dinosaurs help to tell us what they ate. If the dinosaur skull has flat teeth, the dinosaur was a plant eater. If the skull has sharp teeth, the dinosaur was a meat eater.

Activity Book teaching notes

Page 24

- Show learners video clips, posters and books about dinosaurs to help make links between what the dinosaurs looked like and what they ate. Large dinosaurs would find it hard to run, so they ate plants. They would also have different teeth to the sharp, curved, long teeth of the tyrannosaurus, which would be able to tear animal flesh.
- If you feel your learners are ready, introduce the vocabulary *carnivore* (meat eater), *herbivore* (plant eater) and *omnivore* (meat and plant eater), and talk about what these mean. Making links is an important part of thinking scientifically. Compare the dinosaurs with animals alive today that are carnivores, herbivores and omnivores.

Page 25

- Show learners pictures of the skulls of different dinosaurs so that they can compare similarities and differences between them. Talk about how the meat eaters had sharp teeth and the plant eaters had flat teeth. Talk about how the types of teeth were suited to the food that the dinosaurs ate.

- If possible, show learners a picture of a brachiosaurus and an allosaurus before they complete the activity. Talk about the features and habitat of these dinosaurs and what they ate:
  - The brachiosaurus was herbivore – a 'high feeder' like a giraffe today – and it ate leaves from the tops of tall trees up to 9 m in height. Its habitat had lots of tall trees and vegetation.
  - The allosaurus was a carnivore. It had three fingers on each arm to help it tear apart meat. It used to hunt in packs. Its habitat was similar to the brachiosaurus, where it would hunt for prey (its next meal).

Activity ideas

- Ask learners to make play dough or clay dinosaur teeth. Find out about the size of dinosaur teeth and challenge learners to make them the correct size. For example, an allosaurus tooth was around 2–3 cm long but a tyrannosaurus tooth could be up to 30 cm long.
- Place a range of clay 'dinosaur teeth' in the sand tray for learners to excavate.
- Give learners the dimensions of a tyrannosaurus mouth: the jaw could be around 1.2 m long with approximately 50 to 60 teeth that could be up to 30 cm long. Draw this on a large piece of paper on the classroom floor or chalk it outdoors on a hard surface. Learners could lay down inside the mouth to find out how many of them would fit in the dinosaur's jaw. Take photographs to add to a dinosaur display.
- Use *PCM 12: Dinosaur teeth cards*, (on page 96 of this Teacher's Pack). Ask learners to match each tooth to a dinosaur by looking carefully at the size and shape of the teeth. Ask questions such as: *Are all dinosaur teeth the same? What kind of dinosaurs had long sharp teeth? Which teeth belong to a herbivore?*

Success criteria

While completing the activities, assess and record learners who can:
- talk about what different dinosaurs ate
- describe the teeth of different dinosaurs.

How big were the dinosaurs?

 Activity Book C,
pages 26–27

 Story Book C

Learning objectives

- Know that dinosaurs were different sizes.
- Make comparisons between dinosaurs and humans.
- Know what a fossil is.
- Know what fossils can tell us about dinosaurs.

Resources

metre rulers, chalk, string, kitchen or toilet roll paper, clay or salt dough, examples of fossils, pictures of dinosaur fossils, pictures of dinosaur feet and footprints, pictures of dinosaurs and everyday objects, real fossils, tape, natural materials (leaves, twigs, bark, shells)

Key words

measure, length, dinosaur, bigger, smaller, taller, shorter, longer, fossils, teeth, footprints, plants, teeth, shells, ammonites, brachiosaurus (brack-ee-uh-sawr-us), lesothosaurus (leh-soth-uh-sawr-us), velociraptor (veh-loss-ih-rap-tor), allosaurus (al-uh-sawr-us), troodon (tro-uh-don), triceratops (try-sair-uh-tops), stegosaurus (steg-uh-sawr-us), tyrannosaurus (tye-ran-uh-sawr-us)

Background information

Some dinosaurs were huge, like the tyrannosaurus (up to 6 m high), the apatosaurus (up to 27 m long), and the brachiosaurus (neck could reach up to 9 m high). Some dinosaurs were tiny, such as the lesothosaurus (up to 30–50 cm high). Young learners can find it difficult to comprehend the size of dinosaurs so, when talking about their size, it helps to compare them with objects from today, for example: *The lesothosaurus was about the same size as a chicken.*

These activities are designed to help learners appreciate the size of different dinosaurs in relation to themselves. This is an excellent opportunity to link science to maths using early measurement skills. When measuring, learners should use non-standard units, such as sticks, hand spans or cubes, which when counted are within their known number range, for example, up to 20.

We know about fossils because people have found them in rocks. Fossils are the stone remains of animals or plants that lived millions of years ago. Many fossils are the remains of dinosaurs. Palaeontologists find fossils of the bones of dead dinosaurs and dinosaur footprints. Fossils from plants, such as huge ferns that grew when dinosaurs were alive, help us to understand what their habitats were like.

As the dinosaurs walked along, they left footprints. Over time, these footprints filled with mud and sand and eventually became rock. Millions of years later, these footprints become uncovered (through erosion) and this exposes the footprint. Dinosaurs had different-shaped feet. Large plant-eating dinosaurs usually had wide flat feet with short toes to support their immense weight. Meat eaters had narrower claw-like feet with longer toes.

Activity Book teaching notes

Page 26

- Prior to the activity, look at pictures of different dinosaurs and discuss their size. Compare the sizes of dinosaurs to everyday objects.
- In a large indoor or outdoor space, use string or chalk to mark out the length of different dinosaurs. Ask learners to lie in a line to see how many of them will fit into the height of a tyrannosaurus, for example.
- Introduce the vocabulary *taller* and *shorter*. Compare some objects in the class, for example: *I am taller than Sabrina. The tricycle is shorter than Tomaz.*
- Talk about how some dinosaurs were much taller than a human, while others were much shorter.

Activity ideas

- Make comparatives using pictures of dinosaurs and objects. For example, ask: *Was the dinosaur bigger or smaller than our classroom? A giraffe? The school?*
- In a large space, roll out kitchen or toilet roll paper to form the bars of a simple graph to show the heights of different dinosaurs. Place pictures of dinosaurs on the *x*-axis. Engage learners in simple data handling. Ask: *Which dinosaur was the longest? Which dinosaur was the shortest?*
- Use dinosaur facts to have a 'dinosaur Olympics'. Ask learners to find out facts about dinosaurs, for example: *Which dinosaur ran the fastest? Which dinosaur could swim? Which dinosaur could reach the highest tree?* As a class, work out which dinosaur would have won a gold medal (come first) for sprinting, long-distance running, swimming, and so on. Challenge learners to prove they are winners using facts that they have collected.

Fossils

 Activity Book teaching notes

Page 27

- Prior to the activity, talk about fossils – what they are and how they are created.
- Allow learners to handle and talk about some real fossils. Look at pictures of fossils, including pictures of dinosaur fossils.
- Look at pictures of dinosaurs and talk about the size and shape of their feet. Show pictures of dinosaur footprints. You could allow learners to take a print of their own foot to compare with the shape of the dinosaur footprint.
- Talk about how and why dinosaur footprints are still around today.

Activity ideas

- Make a dinosaur's foot in the classroom using tape. Ask learners to place their shoes inside the foot and find out how many of their feet will fit inside.
- Learners could use their own feet to measure the length of a dinosaur's foot.
- If possible borrow or purchase fossils, such as dinosaur teeth or ammonites, for learners to explore and sort.
- Provide pictures and books about fossils, so learners can find out the name of some fossils, for example, ammonites.
- Make fossils using clay or salt dough by pressing a plastic dinosaur into the dough to make an imprint and leaving it to dry. Learners could also press shells or other objects into clay. Take learners outside to collect leaves, twigs and bark and press these into clay to make plant fossils.

Success criteria

While completing the activities, assess and record learners who can:
- talk about the size of different dinosaurs
- make comparisons about the size of dinosaurs
- say what a fossil is
- say how fossils help us to know about dinosaurs.

Dinosaur eggs

 Activity Book C,
pages 28–29

Story Book C

Learning objectives
- Know that dinosaurs laid eggs.
- Talk about the life cycle of a dinosaur.
- Apply knowledge of dinosaurs to make a model.
- Name parts of the body.

Resources

recyclable materials, modelling materials, camera, pictures or video clips of dinosaur life-cycle reconstructions, camera, plastic dinosaurs, play dough, sand

Key words

dinosaur, eggs, babies, grow, adult, nest, model, life cycle, materials, body, footprint

Background information

Dinosaurs laid hard-shelled eggs that differed in size according to the type of dinosaur. Some were laid in nests or mounds, or in hollows in the ground. Once hatched, like animals today, baby dinosaurs were susceptible to becoming prey (being eaten).

There are many similarities between dinosaurs and egg-laying animals today, for example, some dinosaurs incubated their eggs, while others left them to hatch on their own. It is useful to draw on learners' prior knowledge of egg-laying animals, such as chickens, and draw comparisons between these animals and dinosaurs. This helps learners to make links between extinct animals and animals alive today.

 Activity Book teaching notes

Page 28

- Show learners video clips or pictures of what people think a dinosaur life cycle would have been like. Begin with the egg being laid through to the baby dinosaur growing up into an adult and laying eggs, so that learners understand it is a life cycle – it repeats.
- If learners know about the life cycle of other egg-laying animals, make comparisons between these animals and the dinosaurs.

Activity ideas

- Ask learners to use play dough or modelling clay to create a model life cycle of a dinosaur. Photograph the life cycles and create a class book or make a mini video clip describing the life cycle.
- Make dinosaur eggs. Mix sand or soil into play dough to make the egg look more realistic. Press a plastic dinosaur into the centre of the egg and mould into an egg shape. Then the dinosaur can 'hatch' when the egg is broken open.

Design a dinosaur!

 Activity Book teaching notes

Page 29

- In this activity, learners apply what they have learned about dinosaurs to create their own dinosaur model, using various recyclable materials, clay or play dough. They could make a real dinosaur or a hybrid, which could be a mix of their favourites or a totally new dinosaur.
- While learners are making the models, talk about the features of their dinosaurs, for example, how big they are, what they eat, where they live and what their footprints are like.

Activity ideas

- Create a large habitat for the model dinosaurs in the classroom as a display. Invite learners from other classes to visit and learn about dinosaurs.
- Ask learners to share their dinosaur with the class and describe what it is like, how it lives, its habitat, and so on.
- Suggest that learners name their dinosaurs after themselves, for example: *Samosaurus*.
- As a class, create a poem or a song about the dinosaurs.
- Ask learners to role-play the dinosaur that they have created, showing how it moves, roars, and so on.

Success criteria

While completing the activities, assess and record learners who can:

- talk about the life cycle of a dinosaur
- use what they know about dinosaurs to make a model dinosaur and talk about how it lives.

Assessment

- Ask learners to complete the *What can you remember?* activities on pages 30–31 of Activity Book C.
- Ask learners to self-check their understanding of the key learning objectives covered in Units 5 and 6 using the self-assessment chart on page 32 of Activity Book C.

Animals and their babies

owl	owlet	horse	foal
hen	chick	cat	kitten
butterfly	caterpillar	bear	cub
goat	kid	sheep	lamb

Animals hiding

Hodder Cambridge Primary Science Foundation Stage Teacher's Pack © Hodder & Stoughton Ltd 2018

Sea animals

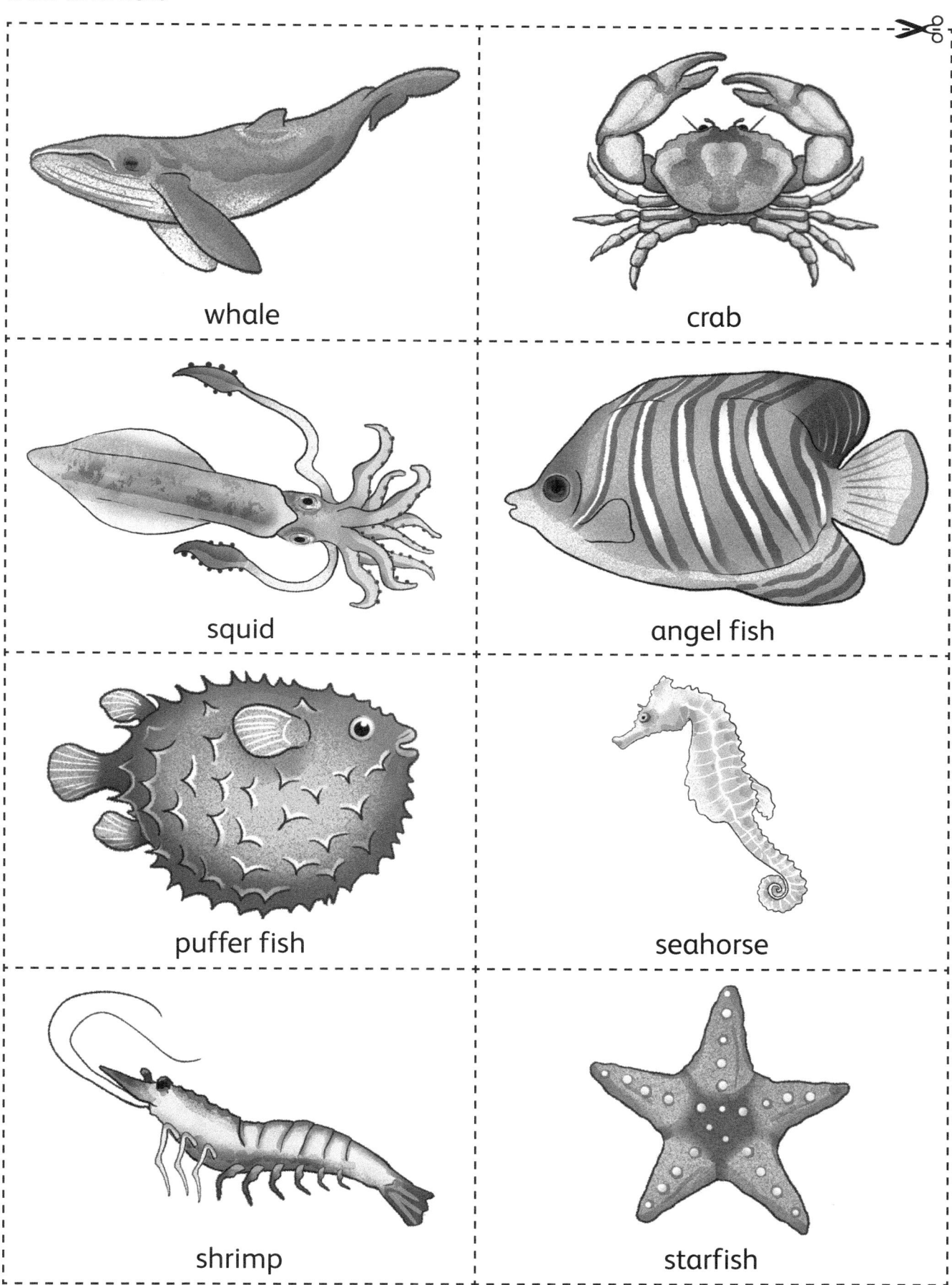

whale

crab

squid

angel fish

puffer fish

seahorse

shrimp

starfish

Sea turtle

Sounds and distance

 Make a sound. Can a friend hear the sound if they walk a few steps away?

 Find out if your friend can hear the sound further away.

 How far away can they hear it? Try out your idea with a friend.

 Draw and write what you did.

Sound-maker cards

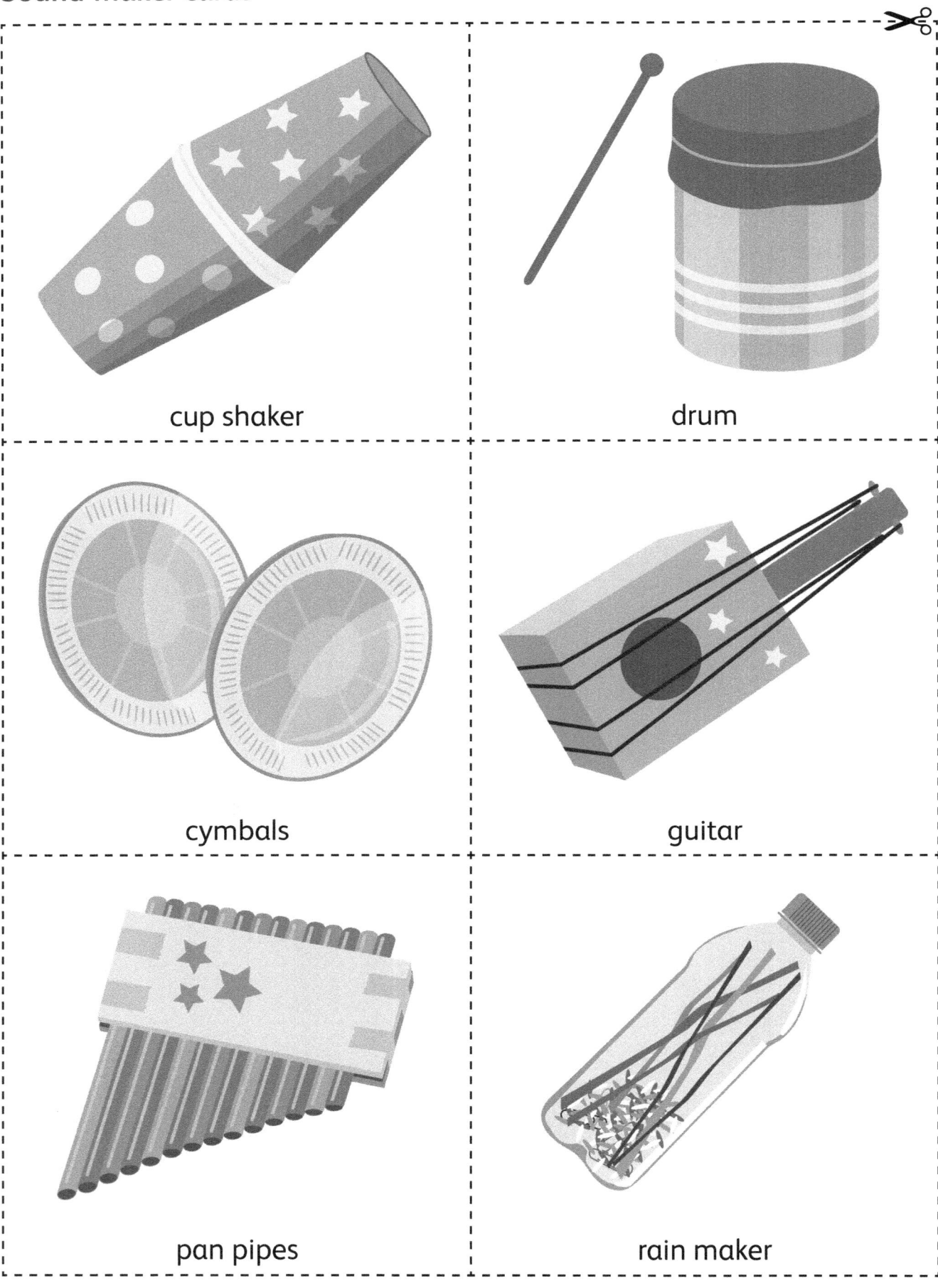

cup shaker

drum

cymbals

guitar

pan pipes

rain maker

Shadow puppets

 Make a shadow puppet. Cut out these shapes. Attach the shapes to wooden sticks. Use the puppets to make shadows.

Construction challenge

My construction challenge was …

Here is a photograph of what I made, I have labelled it.

How it works…

This is what I learned…

Fruit and vegetable cards

apple	pineapple	melon	grapes
coconut	plum	mango	papaya
pumpkin	sweet potato	yam	carrots
spinach	pepper	dates	watermelon
potato	oranges	eggplant	bananas
onions	tomato	cucumber	pear

Sandwich test

 Put a ✓ or a ✗ to show the results of your sandwich test.

I put my bread into …	Is the wrapper waterproof?	Was the bread squashed?	Did it keep the bread fresh?	Will I use it for my picnic?
a plastic box				
a paper bag				
some metal foil				

Dinosaur parts of the body

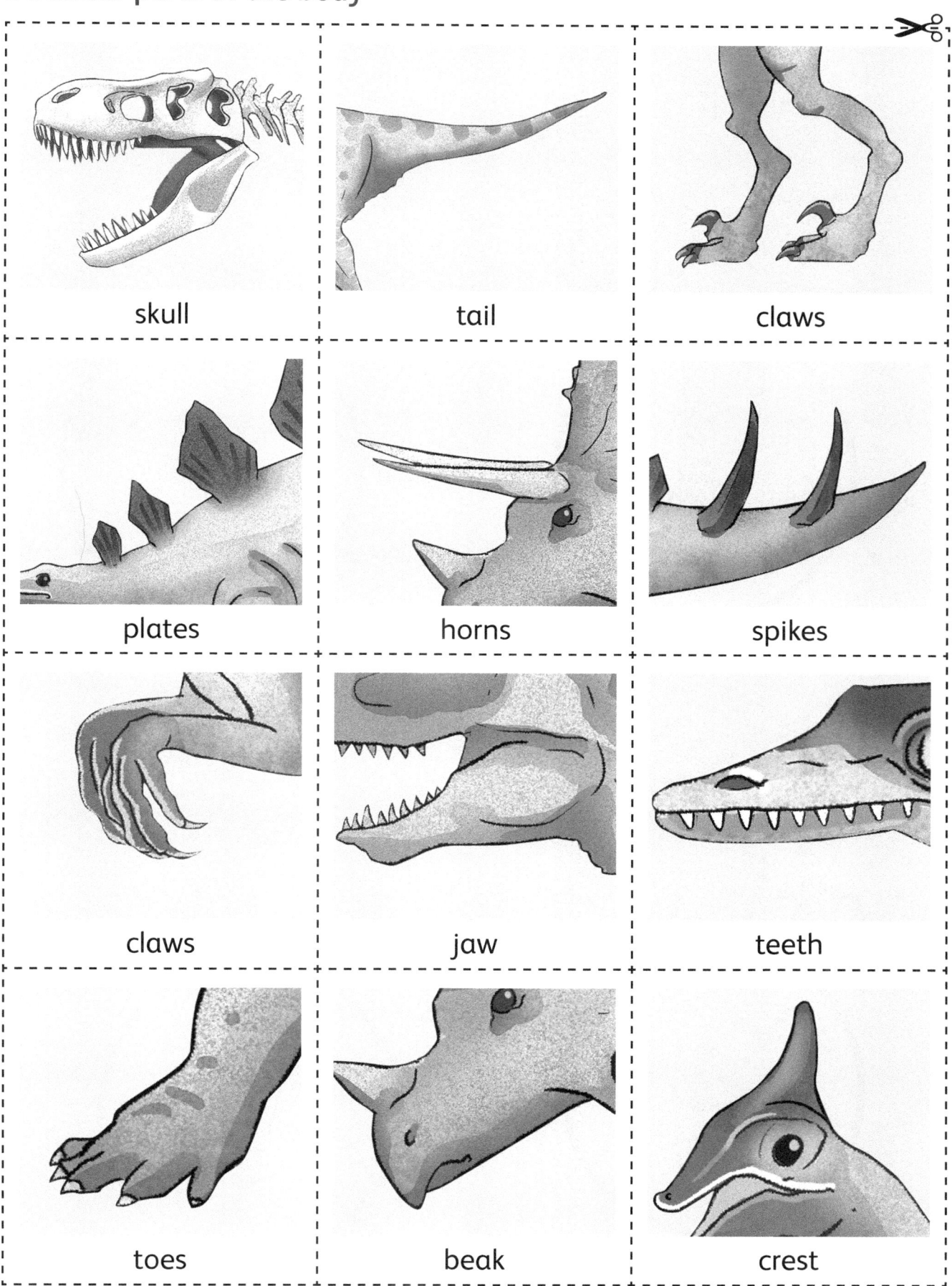

skull	tail	claws
plates	horns	spikes
claws	jaw	teeth
toes	beak	crest

Dinosaur teeth cards

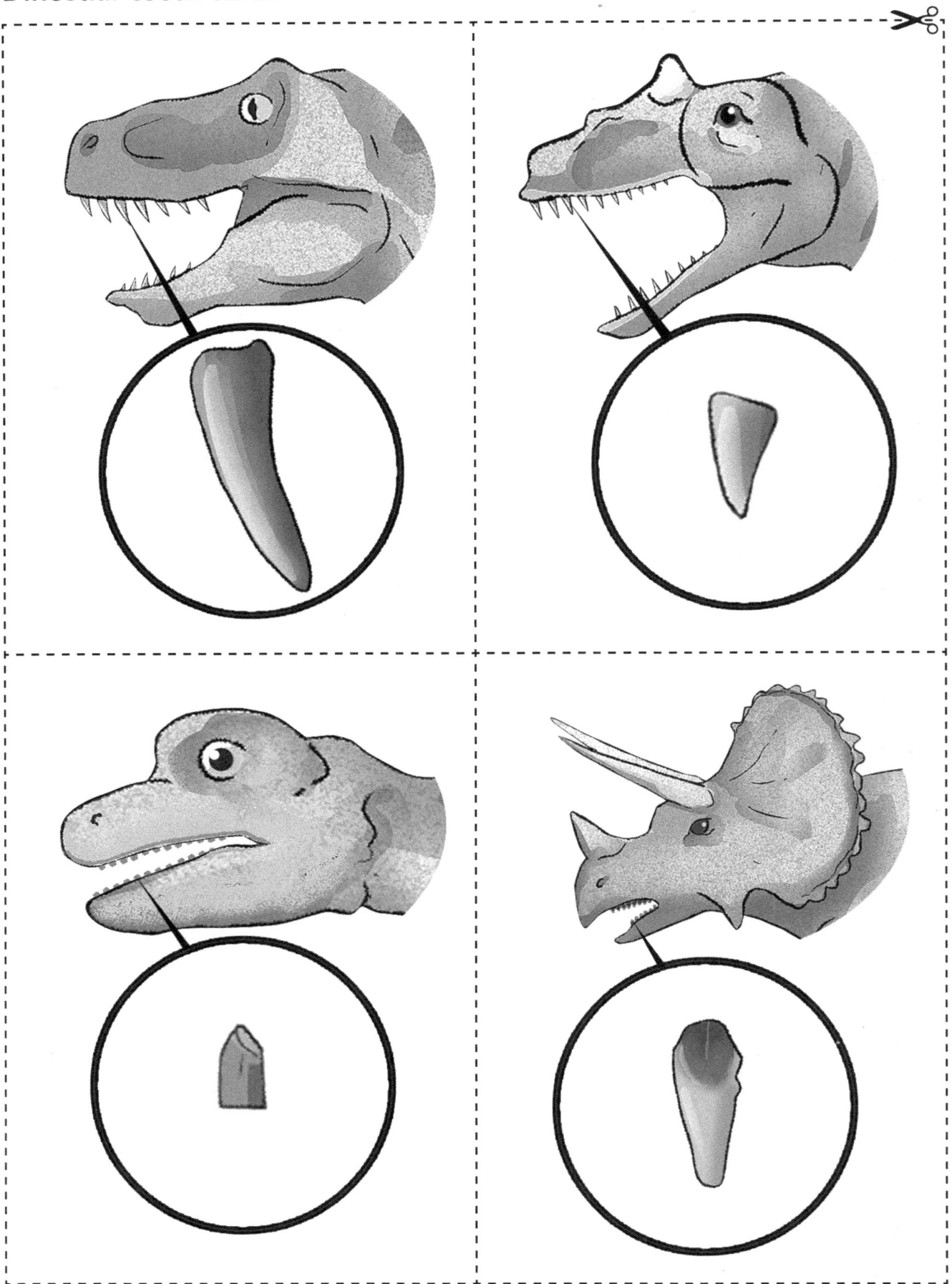